THE EXPERIMENTAL STUDY OF READING

THE EXPERIMENTAL STUDY OF READING

BY

M. D. VERNON, M.A.

CAMBRIDGE

AT THE UNIVERSITY PRESS

1931

CAMBRIDGE
UNIVERSITY PRESS

University Printing House, Cambridge CB2 8BS, United Kingdom

Cambridge University Press is part of the University of Cambridge.

It furthers the University's mission by disseminating knowledge in the pursuit of education, learning and research at the highest international levels of excellence.

www.cambridge.org
Information on this title: www.cambridge.org/9781107455788

First published 1931
First paperback edition 2014

A catalogue record for this publication is available from the British Library

ISBN 978-1-107-45578-8 Paperback

ACKNOWLEDGMENTS

My acknowledgments are due to the Medical Research Council for the grant which made possible the writing of this book. I am greatly indebted to Professor F. C. Bartlett for his help and advice, and for his permission to work in the Cambridge Psychological Laboratory; and to all those others who have from time to time offered many valuable suggestions in connection with this work.

1931 M.D.V.

CONTENTS

Chapter IV

EYE MOVEMENTS IN READING

Chapter V

VISUAL PERCEPTION IN READING

ILLUSTRATIONS

Note. Figs. 6, 10, 11, 12, 13 and 14 are reproduced by permission of the Controller of H.M. Stationery Office.

INTRODUCTION

The purpose of this book is to give a concise account of any experimental work, particularly that recently performed, which throws some light upon the psychology of reading. To understand the nature of the reading processes, it is necessary to have some acquaintance with the psychological and physiological phenomena of vision. Retinal sensation and the various intra-ocular processes integrated with it are fully described in the standard works upon optics, such as Helmholtz's *Treatise on Physiological Optics* and J. H. Parsons's *Introduction to the Theory of Perception*. But the movements of the eye which are mediated by the extra-ocular muscles have been dealt with less fully. Moreover, it has appeared both from the early work of Dodge, Dearborn and Huey, and from the later work from the Education Department of the University of Chicago, that a very remarkable series of specialized ocular motor habits has been developed in the mature reader, comparable to the series of motor habits of the vocal chords in speech and of the hand in writing. These ocular motor habits have been perfected and closely integrated with primary perceptual and assimilatory processes and secondary associative thought processes to constitute the total complex which makes up reading. Consequently, though it has been taken for granted that the reader of this book is sufficiently acquainted with the normal unspecialized phenomena of vision not to require any further description, a full account is given of the rotatory movements of the whole eyeball, and the specialization and development of the habits of movement which occur in reading. To this end, a short summary is first given of the various methods which have been devised for studying and recording these eye movements, followed by a description of the various types of movement mediated by the extra-ocular muscles, and a brief outline of the connections between the kinaesthetic sensations derived from these movements and visual sensation in general. There follows a detailed survey of all the experimental work on the movements of the eyes in reading, and more particularly of the recent work which has not hitherto been comprehensively reviewed.

Having dealt with the motor processes involved in reading, it is then necessary to consider the perceptual phenomena which occur in the reading of adults and children. There has been little experimental work since the publication of Huey's *Psychology and Pedagogy of Reading* upon adult perception in reading; and the majority of the work upon children's perception in reading, though possibly of much pedagogical value, has been too disconnected and uncontrolled to provide results of much reliability or psychological interest. Consequently, these subjects have perhaps been assigned less consideration than they require or deserve. Some attempt has, however, been made to relate them to perception in general. With regard to the secondary thought processes associated with reading we have little experimental evidence; moreover their range is clearly co-extensive with the whole field covered by the higher cognitive processes, and is therefore outside the scope of this book.

Finally, some account is given of the effects of variations in the external stimulus, the printed text. A considerable amount of experimental work has been devoted to this subject, but unfortunately much of it is mutually contradictory, by reason of a series of qualifications and objections which are described therewith. Thus, in the absence of any authoritative pronouncement, our manner of printing continues to be regulated partly by tradition and partly by the cost of printing.

A short résumé of the conclusions as to the development and essential nature of the complex of processes which constitute reading is given at the end of the book.

Chapter I

METHODS OF OBSERVING AND RECORDING EYE MOVEMENTS

(1) Direct Observation

The first attempts at observing and measuring eye movements were made by methods of direct observation. Thus Javal(74) in 1878 observed eye movements during reading by reflection from a mirror. In 1898 Erdmann and Dodge(46) improved this technique by studying the reflected movements through a telescope. These workers were able to determine approximately the number of movements and fixation pauses, but could not measure their duration. Since then the method of direct observation has frequently been used to observe the path of voluntary eye movements and the position of the eye while fixating a point. Thus Barnes(6) in 1905 and Loring (94) in 1915 used very similar methods to measure the angle of torsion of the eye in making oblique movements. The subject of the experiment was required to fixate the object glass of a telescope mounted facing him on a horizontal perimeter, and the cross wires of the eye-piece were focussed upon one of the striae of the iris. The telescope was rotated to another position, the subject's eye following the objective. The angle of rotation of the cross wires necessary to focus them again upon the stria gave the angle of torsion. A more delicate modification of this method was designed by Dr Barany(5), and is described in detail by the writer in another place (145). Öhrwall (109) and Sundberg (140) observed both inter- and intra-fixation movements by means of a Blix's ophthalmometer (13). The eye was illuminated by light from a glow lamp, which passed through one of the microscopes of the ophthalmometer. The other microscope, the eye-piece of which was fitted with a micrometer scale, was focussed upon the eye, and the position of one of the striae of the iris, or of one of the blood-vessels of the conjunctiva, noted on the scale. Movements of the stria, as the eye moved from one position to another, could be measured directly on the scale.

The direct method has been used again in recent years to estimate the number of fixation pauses in reading. In its simplest form, the peep-hole method of Miles (102), a hole about ¼ in. in diameter was made in the middle of the copy to be read. The experimenter held the copy straight in front of the reader, with his own eye close up to the hole. In this way he could gauge the direction of the reader's line of regard, and with a little practice could count the movements of slow readers and children, without distracting the reader. A more elaborate method was devised by Newhall (105) (see Fig. 1). The reader's head was at *S*, where he gripped the horizontal bar with his teeth, and read the copy supported vertically at *R*. The telescope *T*, containing a 20 diopter convex lens at *L*, was trained on the reader's right eye by moving the carrier *C* on a universal joint (underneath the tube) attached to the fixed board *B*. The distance of the experimenter's eye from the lens was just less than the focal length, so that a virtual erect magnified image of the reader's eye was obtained by focussing the lens correctly. The lamp *M* was used simply to produce a general illumination of the reader's eye; it was out of the reader's field of view. For counting the eye movements, a vein or the outer edge of the iris was observed.

Early attempts at measuring the speed of voluntary eye movements were also made by Lamansky (90) and Guillery (63), by the after-image method.

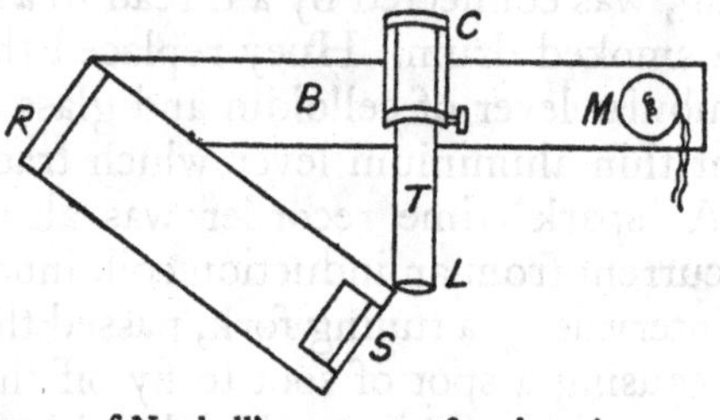

Fig. 1. Diagram of Newhall's apparatus for observing eye movements.

Lamansky found the number of after-images which were produced with intermittent illumination as the eye moved across the field. This method was also utilized by Erdmann and Dodge (46). But it relies too much upon the introspective report of the subject to be very satisfactory.

(2) Devices Attached to the Eye

The first systematic attempt at measuring the speed as well as the number of eye movements in reading was made by Huey (70), using the method invented by Ahrens (2) and adapted by Delabarre (33). Ahrens attempted to record eye movements by attaching to the cornea a light ivory cup, to the apex of which was fastened a bristle which traced on a smoked drum. Delabarre made a thin plaster-of-paris cast of an artificial eye, and trimmed it to the shape of the observer's cornea. The eyelids were propped open, the cornea anaesthetized, and the plaster cast fitted over it. A hole had been drilled in the centre of the cast through which the observer could see to read. A piece of wire, embedded in the plaster with one end projecting, was connected by a thread to a lever tracing on a smoked drum. Huey replaced the wire by a light tubular lever of celloidin and glass connected to a light thin aluminium lever which traced on the drum. A 'spark' time recorder was also used; an electric current from an induction coil, interrupted at regular intervals by a tuning fork, passed through the pointer, causing a spot of soot to fly off the drum at each interruption. This method has been fully described, because Huey made considerable use of it to study eye movements in reading. He claimed that it did no harm to the eye, but acknowledged that the latter frequently felt strained for a time after the cast had been removed.

Orchansky (110), and Marx and Trendelenburg (98), used light metal capsules for the eye. These had small mirrors fastened to them which reflected light from a bright source. In the work of Marx and Trendelenburg the reflected light was focussed on to a strip of bromide paper driven by a clockwork kymograph. By this method even the smallest involuntary tremors could be recorded. It seems to have been delicate and accurate. The experimental period, while the capsule was in position on the eye, did not last more than half a minute. Wiedersheim(155) also employed this method. Struycken(139) is reported to have photographed a minute sphere, mounted on three legs which were attached by hooks to the cornea. It is clear that these capsule methods are all open to the objection that they may interfere with the normal movements of the eye. The drag on the eyeball and the cocainization of the cornea are liable to retard movement. Moreover there is always a danger of some permanent injury to the eye. Consequently Judd, McAllister and Steele(77) evolved a method in which a very much smaller and less dangerous object was attached to the cornea. This consisted of a small flake of Chinese white paint, specially prepared, and attached to the cornea slightly below and on the nasal side of the pupil. It was fairly easily applied and manipulated into the right position, and once there remained firm; if by mistake it was lost under the lid, it was quickly dissolved and did no harm. The movements of this white spot, illuminated by daylight, were recorded by a kinematographic camera, driven at first by hand, but in later experiments by a motor. A time marker connected to the driving handle traced on a kymograph. Later a double camera was constructed with two films and shutters; the exposure of one coincided with the closure of the other, so that the record was continuous. The position of the head was fixed as far as possible by the use of head and teeth grips. Any head movements not eliminated were recorded by the reflection of light from a small steel ball, or a small concave mirror, fastened to a wire spectacle frame.

The chief defect of this method was the immense labour involved in determining the path of the eye movements from the films. For each section of the film the position of the white spot had to be referred to the point of reference provided by the spectacle frame, and the distance measured and recorded on a chart. It was, however, used extensively by these workers for measuring the direction and duration of intra-fixation movements, movements made in viewing optical illusions, and movements of convergence and divergence. The same method has since been used by Totten (144) for recording the eye movements of rabbits.

(3) Photographic Recording

Judd (77) claimed that his method gave a more accurate representation of the path of the eye movements than the purely photographic method of Dodge and Cline (37). But by a method which he has described in detail, Dodge (41) proved that his photographic method was equally accurate. Though the technique is somewhat difficult, this method is so much more convenient and reliable than any other that it has been employed, with various modifications, by all the later workers on eye movements in reading.

Dodge's method (37) was an improvement on one originally used by Stratton (137, 138) to record the movements of the eyes in viewing diagrams. Since in the latter the movements were all photographed upon a single fixed plate, the field was kept in darkness, except for the diagram to be viewed. Moreover, only the number and direction of the movements could be recorded. Dodge and Cline(37) modified this method to measure the angular velocity of eye

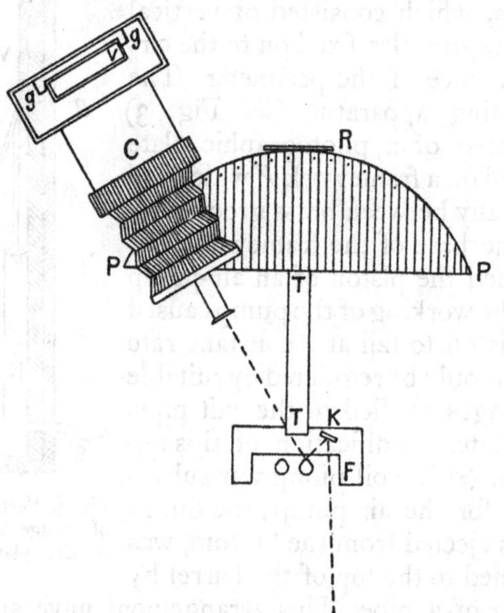

Fig. 2. Horizontal section of Dodge and Cline's apparatus for photographing eye movements.

movements by recording them upon a moving photographic plate. A small piece of white cardboard *K* (see Fig. 2) was illuminated by light from a window behind the observer; and the reflected image of *K* from the cornea, called the 'bright spot', fell upon the horizontal slit of the recording camera *C*, which could be moved round the perimeter *PP* until focussed in the right position. The observer's head was kept in position by a wooden chin-rest *F*; and he was instructed to look through a tube *TT* at the fixation marks, which consisted of vertical knitting-needles fixed on to the circumference of the perimeter. The recording apparatus (see Fig. 3) consisted of a photographic plate carried on a framework *V* which fell vertically between brass grooves *gg*. To the base of the framework was attached the piston of an air-pump *R*. The working of the pump caused the piston to fall at a constant rate which could be regulated by suitable stoppages applied to the exit pipe. In a later modification of this apparatus(38) an oil-pump was substituted for the air-pump; the oil, as it was ejected from the bottom, was returned to the top of the barrel by means of a pipe. This arrangement gave smoother working.

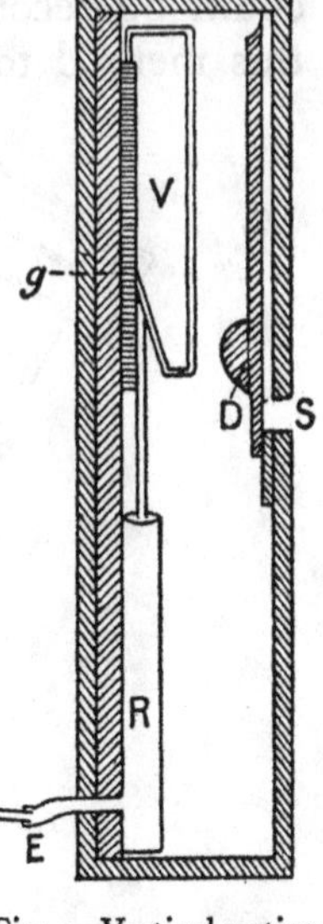

Fig. 3. Vertical section of Dodge and Cline's moving-plate camera.

Dearborn(32), and later Dodge himself(41), studying the movements of the eyes in reading, used an arc lamp as the source of illumination. The rays were rendered parallel by intercepting lenses, and were stopped down to a suitable brightness. Dodge used an alternating current arc lamp giving a series of discrete flashes at regular intervals, and thus obviated the use of a time marker. Piltz (119), Coburn (25), Koch (83) and Weiss (152) all used some variant of the photographic method for recording eye movements, but not the movements made in reading. Their methods are fully described by the writer (145) elsewhere.

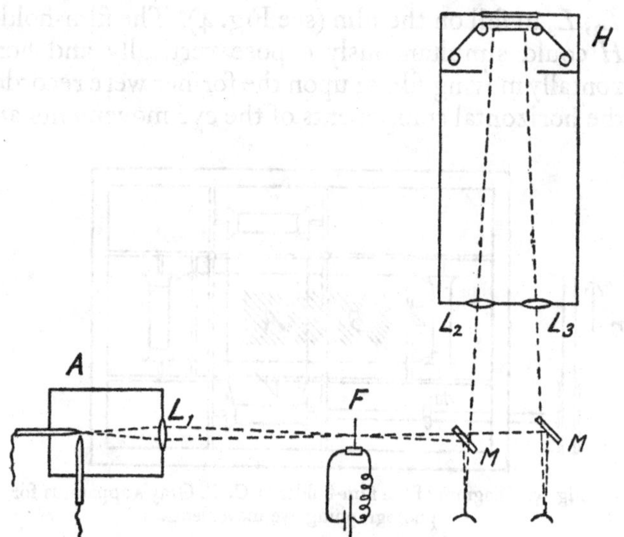

Fig. 4. Diagram of Schmidt's apparatus for photographing eye movements.

The photographic method was used on a large scale for recording eye movements in reading by Schmidt (128), C. T. Gray (59), Judd (81, 82) and Buswell (19). Schmidt's method is shown diagrammatically in Fig. 4. Light from an arc lamp *A* was focussed by a lens L_1 to a point, at which was interposed a tuning-fork interrupter *F* vibrating at a rate of 25 or

50 per second. The light was thus transmitted in discrete flashes, 25 or 50 per second. Later a 400-watt nitrogen-filled bulb was used as the source of illumination. The light was then reflected by thin glass mirrors *M*, *M* on to the corneas of the eyes, and back through the mirrors to the camera; and then focussed by lenses L_2, L_3 to fall on the film (see Fig. 4). The film-holder *H* could simultaneously expose vertically and horizontally moving films; upon the former were recorded the horizontal components of the eye movements and

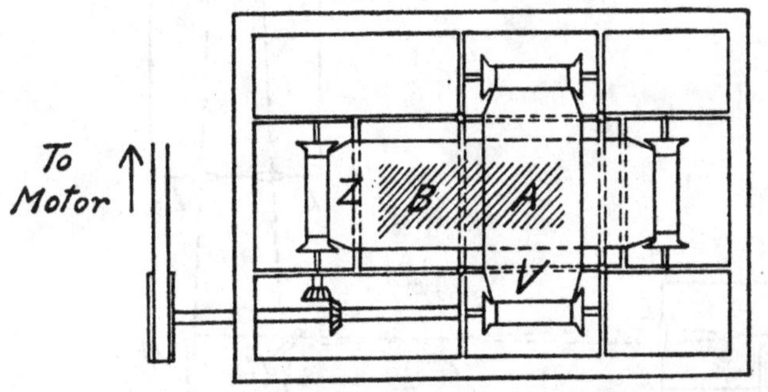

Fig. 5. Diagram of the film-holder of C. T. Gray's apparatus for photographing eye movements.

upon the latter the vertical components. In C.T. Gray's method, the light was rendered parallel by a double convex lens, before reflection from the mirrors. The film holder of C. T. Gray's camera is shown from in front in Fig. 5. Light passed through the focussing lenses to fall on the vertically moving film *V* as at *A*, and on the horizontally moving film *Z* as at *B*. The camera lenses could be focussed simultaneously or separately, and the images of the bright spots on the corneas could be observed upon the ground-glass screen, covering an area equal to $A+B$ at the back of the camera, until just before the films were exposed. The head was secured in a head-rest; any head movements which could not be eliminated were recorded by the method of Judd, McAllister and Steele(77). The reading matter was projected from a lantern placed just above the camera on to a plaster-of-paris surface just below the camera. Fixation points for the orientation of the record, corresponding to the beginning and the end of the lines of print, could also be projected on to this surface. The successive fixations in reading a line of print appeared on the vertically driven film as a series of fine parallel lines composed of dots, each dot corresponding to a period of $\frac{1}{25}$th or $\frac{1}{50}$th of a second. The movements appeared as fine lines connecting the end of one fixation mark to the beginning of the next. The film record could be magnified and projected for measurement to take place.

The only apparent drawback to this method is that the light, although parallel and reduced in intensity by blue glass filters, impinged on the reader's eye directly from in front, and this may have affected his vision to some extent. In the method of Miles and Shen (101) the light was directed by mirrors from beneath. But as the reader was obliged to look downwards to some extent in order to see the reading matter, the light must still have entered the eye at a small angle with the line of vision, and have impinged centrally on the retina. Another modification introduced by Miles was the use of a toothed wheel as an interrupter of the beam of light. This rotated at a fixed rate, and cut off the beam of light each time a tooth of the wheel passed in front of the source.

A further modification was designed by the writer (148) with the object of directing the beam of light more obliquely on to the cornea; it is shown in Fig 6. Light from a 100 C.P. Pointolite lamp situated on the right of the subject passed horizontally through a blue filter and was reflected by a totally reflecting prism P_1 obliquely upwards to strike on the cornea at an angle of about 50°. The reflected beam passed upwards from the cornea to another totally reflecting prism P_2 which directed it into the tube *T* of the camera *C*. The lens *L* of the camera then focussed it on to a film *F* moving vertically in the camera box and driven by a motor. The reader's head was held in position by a head-rest *H* which could be adjusted to bring his eye to the correct position. The image could be focussed, by sliding the lens *L* along the tube *T*, upon a ground-glass screen, which could be fixed in the position of the camera box. A time record was also photographed upon the film by directing a beam of light, periodically interrupted by a time marker, through a hole *U* in the tube on to a small glass mirror *M*. The reading matter was placed opposite to the reader at *R*, slightly below the level of his eyes. This apparatus has not all the elaborate conveniences of those of Gray and Miles; but has the advantage that the recording beam of light only falls on the extreme periphery of the retina when the reader's head is in position. Moreover, since the beam is not itself interrupted, there is no danger of flicker.

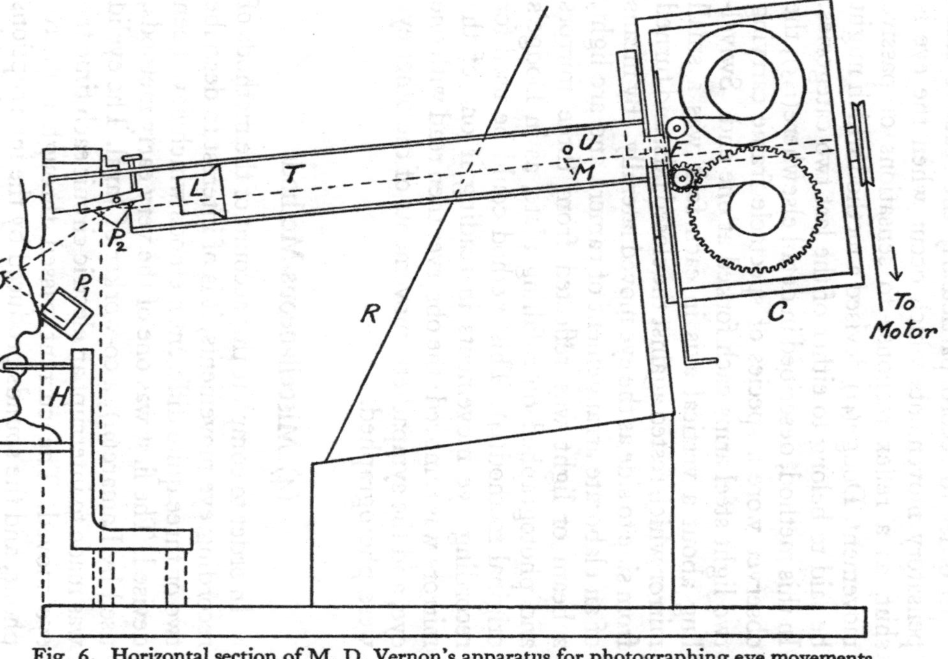

Fig. 6. Horizontal section of M. D. Vernon's apparatus for photographing eye movements.

In order to record particularly the reactive compensatory movements which occur when the eye is shut, as a reflex response to sensations of passive movement, Dodge(42) devised a method which might be said to belong to either of the last two categories. In this method, described in detail elsewhere(145), the observer wore a species of spectacle frame carrying two light steel arms each forked at one end. Swivelling about a vertical axis in each fork was a small mirror which rested against the closed lids, and turned from side to side as the eye moved laterally. By means of an elaborate arrangement of camera and arc light, a beam of light was reflected from three mirrors and photographed on a falling plate, as in Dodge's original method(44). This method could be used for recording eye movements in reading if one of the mirrors was removed; the observer then read with one eye, and the sympathetic movements of the other eye were photographed.

(4) Miscellaneous Methods

In order to complete this account of the methods of recording eye movements, it is of interest to describe two or three quite different methods which have been devised. The first was one of the very early methods, used by Lamare(91), a co-worker of Javal. The eyelid was made to interrupt an electric circuit each time the eye moved. The current passed through a microphone, and the sounds produced by the interruptions were counted. This method seems to have been inadequate, however, and has not been repeated. Another method was that of Schackwitz(127). A small distended rubber capsule was mounted upon an adjustable spectacle frame in such a manner that it rested upon the upper eyelid. It was connected by a brass tube and a rubber tube with a Marey tambour. Movements of the eyelid, and of the eyeball beneath it, were recorded by variations in the compression of the rubber capsule, which caused the diaphragm of the tambour to vibrate. Galley(51) devised an elaborate system of air reservoirs and tubes, by which

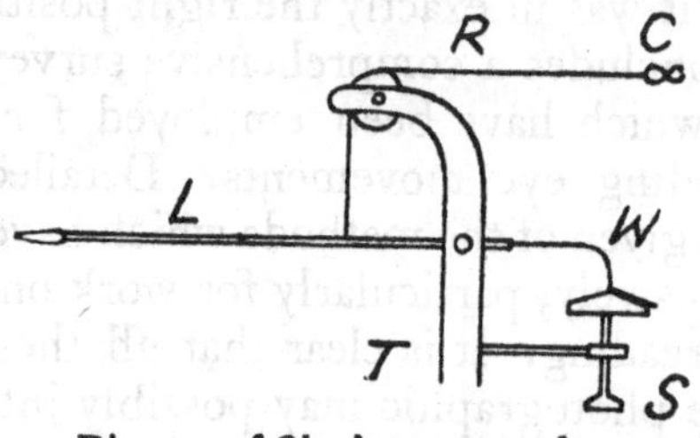

Fig. 7. Diagram of Ohm's apparatus for recording eye movements.

movements of the diaphragm of the Marey tambour were transmitted, greatly magnified, to another tambour which regulated the tracing of a lever on a drum. Galley claimed that he could thus record eye movements in reading; clearly, however, there is a danger of confusion with purely lid movements.

Another method was that of the lever system, invented and much modified by Ohm, and used by him, and by Engelking(45) and Cords(27), to record nystagmoid eye movements. Much work was done by Ohm(107) on miners' nystagmus with this method. The latest modification(108) is shown in Fig. 7.

The small clip *C* was fastened to the cornea, laterally to the pupil. As the eye moved from side to side, movements were transmitted by the thread *R* to the lever *L*. By raising or lowering the screwhead *S*, the tension on the short arm of the lever was adjusted, by means of the watch-spring *W*, until it counterpoised the weight of the long arm of the lever, and there was no pull on the eye. The support *T* was connected to the chin- and head-rest, and could be adjusted in three directions at right angles to each other, till it was in exactly the right position.

This concludes a comprehensive survey of all the methods which have been employed for observing and recording eye movements. Detailed accounts have been given of the methods which have been used most extensively, particularly for work on eye movements in reading. It is clear that all these methods except the photographic may possibly interfere with the normal movements of the eye; and even in photographic registration, too strong a beam of light may affect them to some extent if directed centrally on to the retina. If this provision be borne in mind, however, the photographic method provides an accurate and delicate method of recording and measuring eye movements and fixation pauses.

Chapter II

TYPES OF EYE MOVEMENT

(1) Simple Reactive and Sweeping Eye Movements

Several different types of rotatory movement of the eyeball are produced by the action of the extra-ocular muscles. The best known, and also the most important from the point of view of reading, are reactions to eccentric retinal stimulation. When the image of an external object falls upon the periphery of the retina, the eyeball is rotated reflexly until the image falls upon the fovea and clear vision is obtained. Voluntary eye movements from one point to another and sweeping movements for surveying the visual field are similar in character; so also are the movements of the eyes in reading. But even when initiated voluntarily, these movements are typically reflex in character during their execution. Their speed cannot be varied voluntarily, nor can their direction be changed except by pausing and then moving in another direction. There seems to be little doubt that no perception takes place during movement. Thus if one watches one's own eyes in a mirror, it is impossible to see one's own eye movements; which seems to show that one cannot perceive the field of vision during movement. Holt(68) showed that no perception occurred during voluntary movement, since after-images could be generated which were perceived by the eye when it was at rest but not when it was in motion. And also a stimulus which appeared while the eye was moving was not perceived till it had come to rest. He attributed this to central visual anaesthesia during eye movement. Dodge(40) considered that we are unaware of any perception during motion, but that this is not due necessarily to actual central visual anaesthesia. This view is supported by Leiri(92), who found that in certain cases eye movements were accompanied by peripheral, though not by foveal, vision (in the dark-adapted eye). He concluded that normally the vague blurred visual impressions obtained during movement are prevented from reaching consciousness by the strong clear foveal impressions received during fixation. Certain it is that, from whatever cause, we are unconscious of the fusion of sensations which would necessarily occur as the result of the rapidity of the eye movements, as Dearborn(32) pointed out. This may be due merely to the fact that we have learnt to ignore these blurred sensations, as we ignore the blind spot, and various entoptic phenomena.

In general the movements of the two eyes are co-ordinated, and one cannot move voluntarily without the other; this co-ordination is not present at birth, but develops soon afterwards. The movements, however, may not always be identical, either in their path or their velocity. Indeed, Judd and Courten (76),[1] photographing the movements of the eyes in viewing the Zöllner illusion, found that the eyes sometimes moved in directions opposite to each other (not in movements of convergence and divergence). This

[1] Experimental method described on p. 5.

is probably due to a difference of muscle balance between the two eyes; it varies in different individuals. Kolen (87) has cited cases of individuals who could voluntarily perform movements of one eye while the other remained stationary. He stated that these were not due to abnormalities in the innervation or musculature of one eye, but to a new conditioned reflex, and that they could be learnt and perfected with practice. Bramwell (16) denies that such movements can ever take place; and no doubt they are very unusual.

Owing to the fact that no perception takes place during movement, it is clear that the extent of a movement cannot be adjusted during its execution. Consequently, in making unusual movements, especially those of large angle, the eye very frequently, if not almost universally, stops at some position near the correct point, and then makes one or more short and usually jerky corrective movements until it reaches that point. Dearborn (31)[1] found that the mean extent of these corrective jerks in moving horizontally to a point 40° from the primary fixation point was 1° 48′ for one subject, and 2° 59′ for another; but after the formation of 'short-lived motor habits' with practice in making this particular movement, the mean extent was reduced to 1° 11′ and 1° 42′ respectively. With an angle of movement of 10°, McAllister (95)[2] found at the first trial a corrective movement of 1° 18′; the corrective movements became much smaller with practice, but were never entirely

[1] Using the experimental method of Dodge and Cline (37) (see p. 8).

[2] Experimental method described on p. 5.

absent, or exactly alike on two occasions. Moreover, they were different for the two eyes at the same movement. It is clear that some special motor ability is involved, which is subject to a temporary improvement with practice in one particular movement, but may also receive a more permanent facilitation. Thus Stratton(138) photographed the movements of the eyes in following the lines of a diagram, and found that horizontal movements were more accurate, and less liable to corrective glides and angular changes of direction, than vertical and oblique movements, since these are much less frequently utilized. It was found by Miles(103)[1] that in making movements between two widely separated points the eye does not as a rule move far enough at first, although occasionally it moves too far. The conclusions of Miles are supported by the writer, who found (149)[2] that in making voluntary horizontal movements of 5° to 40° there was a preponderating tendency to fall short of the correct point, increasing with the angle of movement; but a certain number of movements beyond the correct point did occur. The average values for nine observers are given in Table I. From these figures it appears that movements from the centre to the periphery were considerably less accurate than equiangular movements symmetrical about the central position, and were only slightly more accurate than movements of double the angle which were symmetrical about the central position. Thus apparently the accuracy of the movement depends upon the dis-

[1] Experimental method described on p. 11.

[2] Experimental method described on p. 12.

tance of the fixation points from the central position, rather than upon the angle of movement. It was found that individuals varied considerably both in the general accuracy of their movements and in their tendency to move either too far or not far enough. Thus altogether 89 per cent. of the movements of one observer were inaccurate; 80 per cent. were too short and 9 per cent. were too long. On the other hand, only

Table I.

Angle of movement	Angular distance of fixation points from central position	Percentage of movements short of correct point	Average angular distance short	Percentage of movements beyond correct point	Average angular distance beyond
5°	5°	51	47′	16	26′
10°	5°	52	56′	18	30′
	10°	65	51′	14	39′
20°	10°	70	1° 17′	14	34′
	20°	80	1° 26′	9	51′
40°	20°	76	2° 4′	10	32′

62 per cent. of the movements of another observer were inaccurate, and of these 46 per cent. were too short and 16 per cent. were too long. The other observers varied between these. Clearly the greatest source of inaccuracy was the inability to go far enough in a single movement without stopping to see how far the eye had gone. It seems probable that these preliminary pauses were made to obtain a clearer view of the desired fixation point in order to move on to the right position. It was shown by Dodge(41) that a latent period was necessary between one fixation and

the next, in order that the after-effects of the previous stimulation might clear off and the new stimulation might develop. This latent period he termed the 'clearing up period'; an impression was practically 'cleared up' when it could be differentiated from its predecessor. The clearing up process took place in part during the eye movement, but a certain additional interval of time was necessary after the beginning of the fixation before the impression was fully cleared up. The duration of the interval varied with the nature and illumination of the field, the size of the object, and the position of stimulation of the retina; but under conditions such as those which obtained in looking backwards and forwards between two points, it was probably not less than 100σ. Now Miles(103) found that the duration of the preliminary fixations in moving from point to point was about 150σ, while in the writer's observations(149) it was frequently as much as 200σ. Thus it is clear that these fixations were of sufficient duration to allow the incubation of a stimulation to corrective movements in response to retinal sensation, although, as already mentioned, the latter was never present in consciousness.

The writer found that if the first fixation of the eye after movement was situated fairly close to the correct point, frequently no jerky corrective movement occurred, but the eye glided gradually to the correct position. Occasionally it would remain throughout the period of fixation in its first and incorrect position. Miles(103) found that when the observer was not co-operating well in the experimental procedure the final fixation at the outer position (i.e. away from the nose) was less accurately adjusted, and that corrective jerks were sometimes omitted. With even less co-operation, or when the alternation of movements and pauses was very rapid, corrective movements were sometimes omitted at both the inner and outer fixation positions.

These observations, together with those made by Stratton (137) on the movements of the eyes in viewing figures and diagrams, show that inaccuracy of movement is not merely a motor phenomenon, due to lack of muscle balance or of properly adapted motor ability. Stratton (137) found that when an observer was instructed to follow the lines of a simple diagram with his eyes, these movements were jerky, irregular and inaccurate. In other directions than the horizontal they were frequently wrongly directed and tended to swing in towards the nasal side, as the result of overaction of the internal rectus (138). Curves were followed less accurately than straight lines. In observing symmetrical objects the movements were not symmetrical, and the fixation points were situated asymmetrically. These inaccuracies were no doubt partly due to motor inability, but partly, in all probability, to what R. MacDougall (96) called the visual 'reflex attraction' of objects in the peripheral field. MacDougall required his observers to estimate the median point between two cardboard rectangles placed in a horizontal line, one of which remained constant while the other varied in size, form, brightness, or colour. He found that a reflex attraction was set up by objects of greater light intensity than those seen centrally, and by novel and unusual objects in the periphery. He also demonstrated the habit of the eye, when exploring a system of visual objects in a limited field, of coming to rest in a position where there was equilibrium between stimulations to reflex movements of rotation in different directions. It was pointed out that the primary function of these movements was to bring to the fovea the images of objects seen peripherally. It is obvious, then, that in viewing any complex figure there will be a tendency to obtain as rapidly as possible foveal impressions of all parts of interest which make up the figure. Hence it may be deduced that it is difficult to execute movements according to any regular and well-ordered plan, and that an effort of attention is required to prevent the eyes from oscillating rapidly to and fro between all the salient points of the field of vision, which form of behaviour is typical of a vague and general survey of the field. That this tendency is only partially suppressed is shown by the frequent irregularity and inaccuracy of the voluntary movements.

It was found by Dodge(37)[1] that the velocity of the eye in moving horizontally between a series of points varied with the original position of the eye and the direction and angle of movement. It was also affected by fatigue and individual variations, but was constant for the same individual when the above conditions remained constant. The values given by Dodge for lateral movements may be compared with those found by Koch(83), using a method similar to that of Dodge and Cline (see Table II). It will be seen that these results agree fairly well. They both show that the rate of movement increases with the

[1] Experimental method described on p. 7.

angle of movement. Dodge concluded that each movement consisted of three phases, the first of acceleration to the maximum velocity which constituted the second, followed by the third, of retardation. Thus in short movements there would be insufficient time to accelerate to a high velocity. Hence also the very high value observed by Koch for the maximum velocity in the middle of long movements.

Table II.

Dodge's results		Koch's results	
Angle of movement	Angular velocity per second	Type of movement	Angular velocity per second
5°	174°	Slight tremors	50°
10°	258°	Corrective movements at fixation pauses; Smaller voluntary movements	100°–200°
15°	311°		
20°	365°		
30°	373°	Larger voluntary movements	200°–500°
40°	400°	Maximum speed in middle of movement	700°

The values for the rate of movement in reading have been found to be very similar to those for voluntary and reflex movements. For the forward, inter-fixation movements of about 4°, Dodge and Cline(37) found an average velocity of about 200° per second, Huey(72)[1] about 80° per second, and Dearborn(32)[2] about 110° per second. For the return sweeps of 10° to 14° from the end of one line

[1] Experimental method described on p. 4.
[2] Experimental method described on p. 8.

to the beginning of the next, Dodge found an average velocity of about 320° per second, Huey about 230° per second, and Dearborn about 250° per second. There is very little doubt that the eye movements of Huey's observers were retarded by the plaster-of-paris cup placed on the cornea. These values are all, however, of the same order as those shown in Table II. The increased speed for the longer movements accounts for Huey's observation that the return sweeps lasted but little longer than the forward movements.

It was observed by Miles (103) that abductive lateral movements are slower than adductive movements, that is, the rate of movement inwards towards the nose is greater than the rate of movement outwards away from it. This may account for the observation of Koch (83), that the rate of movement of the two eyes sometimes differed considerably, since in lateral movements one eye would be moving outwards, the other inwards. Stratton's observations appeared to show that oblique and vertical movements are slower than lateral ones. Dearborn (32) also found that the velocity was rapidly decreased by fatigue in reading, particularly the velocity of the long return movements from the end of one line to the beginning of the next. These observations were, however, made only once on one reader, namely himself, after a hard day's work. Miles (103) found that the velocity varied regularly diurnally from 90σ to 140σ for a movement of 40° (from about 280° to 450° per second); the mode was 100σ to 110σ (360° to 400° per second). The speed decreased towards the end of the day, and became very slow at the onset of sleep, e.g. 250σ for a 40° movement (160° per second). Also the pause before the corrective shift became very short, as if vision occurred during movement, and hence corrective movements were scarcely necessary to reach the correct position. Finally these movements were replaced by rolling circular movements. Thus it seems that as sleep supervenes, the normal reactive movements are replaced by another type of movement similar to the co-ordinate or reactive compensatory movements to be described below.

The velocity of the eye movements varies to some extent from one individual to another, though possibly not as much as does the accuracy of movement. The variation of average velocity between the fastest and slowest of Dodge's three observers(37) was only 7 per cent. Abnormal cases may show very large variations. Diefendorf and Dodge(35) found that in maniacal states the movements were abnormally rapid, possibly because higher nervous control is lessened; in depressive states they were abnormally slow, because there is a general loss of efficiency of motor activity of all kinds.

(2) Convergent and Divergent Movements

Next in importance to simple reactive movements are convergent and divergent movements, which represent a higher level of co-ordinated movement. They are dependent upon the possession of binocular vision, which only exists in the primates; and in the human child they do not develop fully until about the sixth year. The tendency to make these movements seems to be innate, but is developed as a habitual response. They correspond to the level of 'epicritic' sensibility, while simple lateral movement corresponds to the 'dyscritic' level (Parsons(115)). They provide a co-ordinated reflex response of both eyes to peripheral stimulation, designed to focus the optical images of external objects upon 'corresponding points' in the two retinas. By means of sensations from these corresponding points a single object is perceived in space; but the slight difference between the two optical images gives the object an appearance of depth and solidity. The convergent and divergent movements are relatively slow, lasting 350σ to 400σ (Judd(79)[1]), as compared with 20σ to 40σ for short lateral movements. Dodge(41) considered that they were slow enough to allow the existence of perception during their execution without fusion of the stimulations. He attributed this slow rate of movement to retardation by the tendency to sympathetic lateral movement of the two eyes. Judd found that a number of fine corrective movements usually took place before convergence was complete. Sometimes a sympathetic lateral movement occurred before convergence began; one eye moved laterally in sympathy with the other and then gradually re-adjusted itself to the convergent position. It was unusual for both eyes to follow paths of the same form, or to proceed with the same rapidity. The dissimilarity was often an individual peculiarity; the left eye might move after the right, or vice versa. This dissimilarity was not dependent upon

[1] Experimental method described on p. 5.

right- or left-eyedness or differences of visual acuity. It was attributed by Judd (79) to differences of muscle balance; for the degree of dissimilarity was altered when the fixation points were changed, that is to say, when the muscular tension of the eyes altered.

Schmidt (128)[1] found that convergent and divergent adjustments took place regularly during reading. Convergence occurred at the beginning of each pause, and divergence during movement; the convergence was greatest at the first pause in each line of print, presumably because there had been time for a large degree of divergence during the long return movement from the end of one line to the beginning of the next. But this form of binocular adjustment only developed fairly late. In immature readers there was often found at the beginning of the fixation pause a rapid lateral movement of both eyes in a direction opposite to that of the normal direction of movement, that is, from right to left instead of from left to right. In the mature and practised reader this type of movement had been superseded by convergent and divergent movements, showing again that the latter represented a higher form of development.

(3) Torsional Movements

All movements of the eyes other than horizontal and vertical are accompanied by torsional movements of the eyeball about a horizontal axis passing through the centre of the eyeball from front to back. The angle of torsion has a definite relation to the position of the

[1] Experimental method described on p. 9.

line of regard with respect to the head; for a given degree of elevation, the angle of torsion increases with the degree of lateral displacement, and for a given degree of lateral displacement, the angle of torsion increases with the degree of elevation. Listing's Law gives the actual relationship. These torsional movements have been evolved in order that the eye may attain the desired direction of the visual axis with the least effort and the least possible tension on the extra-ocular muscular and connective tissue. The connections between the sweeping, convergent and torsional movements of the two eyes, and their relationship to accommodation, are not fixed and absolute or dependent upon anatomical mechanisms. They have developed through use and habit, are liable to variation, and can be altered to some extent volitionally.

(4) Pursuit Movements, Co-ordinate Compensatory Movements and Reactive Compensatory Movements

Three further types of movement have been distinguished and photographed by Dodge(38). Though of no importance in reading, they may be described for the sake of completeness. The first of these are known as pursuit movements, and take place when the line of regard follows an object moving across the field of vision. They consist of long, smooth movements at a rate regulated by that of the moving object, but usually slightly slower, interspersed with short, rapid, jerky corrective movements made in order to catch up with the moving object. The pursuit movements become more accurately adjusted to the movements of the object after a little practice, and persist for a time after the object has ceased to move.

Secondly, there are co-ordinate compensatory movements, by which constant fixation of a stationary object is maintained during rotation of the head. These movements begin as soon as do the head movements; they are steady and accurate throughout, except when the head movements are very large or rapid, and are not interspersed with secondary corrective movements. They also occur when the trunk, or even the legs, are moved. It is found that the eyes cannot remain fixed in the head if the latter is moved, unless following the path of an object with a direction and angular velocity the same as that of the head movement.

Lastly, there are reactive compensatory or nystagmoid movements which occur when the eyes are shut, in reaction to organic sensations of passive movement. These movements are difficult to observe since they occur in typical form only when the eyes are shut; but of late years they have been studied in detail by Dodge by means of the mirror reflector described on p. 14. He found (43) that they occurred as direct reflex responses to vestibular stimulation; they were always in the direction of compensation for the movement of the body, and may have been approximately adequate to compensate for the actual speed. They provided a non-cortical reflex reaction of low latency to sensations of passive movement; but normally they were rapidly superseded by visual pursuit or co-ordinate compensatory movements which are regulated by the cortex and are more delicately adjusted to the process of compensation.

(5) Fixation Pauses

It is now necessary to consider the fixation pauses which occur between one simple reactive movement and the next. The power of reflex fixation is only slight in infants, and their fixations are momentary. Longer fixations develop with age, in response to the demand for a clear foveal perception of the object fixated. The duration of the fixations is considerably longer than that of the reflex movements, which rarely last more than 150σ even for large-angled movements. On the other hand, the fixation pauses are not as a rule less than 150σ, since, as was mentioned on p. 22, time must be allowed for the 'clearing up process', lasting about 100σ, for the effects of previous stimulation to clear off before perception and fresh stimulation can occur. Huey(72) and Dearborn(32) give the minimum duration of the fixation pause in reading as 160σ and 170σ. On the other hand, McAllister(95) found that when taking a kinematographic film, at the rate of nine exposures per second, of the fixation of a point by the eye, only on one occasion was the eye in the same position in two consecutive exposures. Thus absolute fixation could not be maintained for a period of more than 100σ, and frequently was maintained for less. These movements away from the fixation point were of the order of 30′ of arc, and cannot probably be classed with the ordinary inter-fixation movements. It is probable that they are in part muscular tremors dependent upon the relative tension of the various extra-ocular muscles. But Judd(78) considers that their chief cause is the sensory modification of the retinal elements upon which the image of the fixated object is focussed. A change of excitability in the receptor organs and their nervous connections is produced by stimulation; and to restore sensory equilibrium it is necessary to transfer the stimulation to an adjacent area, the sensory response of which is practically identical. Hence the frequent slight shifts and movements occurring during fixation. Now Dodge(41) has shown that for any given object of regard several retinal positions are practically equivalent. A clear after-image of a wedge of light was made to fall on a particular point in an object of regard, such as a printed word, by fixating that object with the eye. The eye was turned away, and then back again and the object re-fixated; and it was found that the after-image fell somewhere else. That is to say, when an object is fixated, the optical image of almost any part of it may fall upon the fovea. Slight displacements of the optical image do not affect the retinal sensations appreciably. Thus there is no tendency to inhibit these shifts, whatever their cause. Dearborn(32) found that these shifts took the form of a slow gradual drift away from the fixation point. This was observed by the writer(149) to occur during fairly long periods of voluntary fixation with some observers, but not with all.

McAllister(95) found that if the point to be fixated had a series of lines radiating from it, the area covered by the intra-fixation movements was greatly increased. This Dodge(39) attributed to the diminution in clearness of the fixation mark. But he found that a single line through the fixation point reduced the amplitude of movement, presumably since it provided a clearer and more visible fixation mark. It seems probable that the effect of the radiating lines is to provide peripheral stimulation which is liable to set up reflex movements of the normal type, described at the beginning of the chapter. It was found by the writer (149) that when an observer was required to fixate a series of small crosses on a screen at angular distances from the central position varying from 5° to 20°, there was a constant tendency for his eyes to drift away from and jerk rapidly back to the fixation cross, or vice versa, as a result of peripheral stimulation by the other crosses. These movements covered angles of $30'$ to $40'$, but the altered fixations had a duration of the order of 200σ. The movements were thus probably reflex responses to peripheral stimulation, and not true intra-fixation movements. Dearborn(32) observed that fixation and movement in reading were indistinguishable with some of his readers; the eye drifted over the print, and perception took place during movement. It is difficult to explain this behaviour, unless it can be attributed to some ocular defect. His observations have not been repeated or confirmed by other investigators.

Chapter III

SENSATIONS OF EYE MOVEMENT

(1) Kinaesthetic Sensations of Reflex Eye Movement

Great claims have been laid by some writers, such as G. T. Stevens in his book on *The Motor Apparatus of the Eyes*(136), to the importance of the kinaesthetic sensations which we obtain from the movements of the eyes. Perception of space and depth, and evaluation of the rate of movement of external objects have all been attributed in whole or in part to sensations of eye movement. It may then be of some interest to examine the truth of these claims, since the types of perception mentioned are of such fundamental importance.

Let us consider first the reactive and sweeping movements. It was stated in the last chapter that we are not conscious of the irregularities and inaccuracies of these movements, nor of the pauses and corrective glides necessary to adjust them. The writer (149) found that the tendency to stop short of the correct point in making voluntary movements was much greater with every observer than the tendency to go too far. Yet two-thirds of the observers reported that they felt a tendency to go beyond the correct point; only one observer reported a tendency to fall short, and he actually stopped short on fewer occasions than any other subject. Stratton (137) found that the eye movements of an observer viewing a diagram were jerky, irregular and inaccurate, though the subject thought they were regular and smooth. This was probably because his perceptual processes moved regularly from point to point, and gave the impression that his eyes did so also. Again, we are not in general aware of the alternate movements and fixation pauses in reading, but feel that our eyes move smoothly backwards and forwards along the printed line. It was found by the writer (146) that observers fixating a series of letters and words, widely spaced in horizontal lines on a distant screen, were unaware of the exact nature of their eye movements. They could only report rather long fixation pauses and long hesitations and regressions; and even these reports were not reliable. It seemed that the observers were usually aware of fixations and re-fixations during which prolonged scrutiny and complete apperception of the fixated object took place, but not of the actual stoppages or reversed movements of the eye, and the altered muscular tensions and strains which these involved. Again (149), one reader was able to estimate the proportion of the printed line covered by his eye, not by means of direct kinaesthetic sensation, but by remembering the first and last points fixated in the lines; that is, by perceptual means.

Some indication of the general ease or difficulty of the eye movements may be obtained from the kinaesthetic sensations. In fixating the series of letters and words mentioned above (146), it was found that sensations of the general facility and regularity of movement were closely related to the objective accuracy of the performance. This was particularly noticeable when the differences of ease and efficiency were marked, as between reading disconnected words and words connected in sentence form. But the ocular motor habits involved in reading connected sentence material are extremely efficient and deeply rooted, and the contrast with the muddled and frustrated kinaesthetic sensations which occurred while reading the disconnected words was great. Again (149), a feeling of general fatigue of the eyes sometimes was paralleled by a decreasing accuracy of voluntary movement from one point to another. When there were feelings of strain and tension in the extra-ocular muscles only, no increase of irregularity of movement or fluctuation of fixation could be detected. In any case, it is clear that such general sensations of ease, regularity, strain or fatigue afford little aid to perception. It is not even likely that a wide-angled space is estimated from the tension set up by wide-angled movements, since the natural procedure in such a case would be to move the head from side to side, not merely the eyes in the head. A measurement from the combined head and eye movements would be extremely inaccurate compared with the fine degree of retinal discrimination.

(2) Spatial Perception

There does, however, seem to be some direct and unconscious relationship between the actual movements of the eyes and the accompanying retinal sensations. Thus Judd (78)[1] found that in viewing the Müller-Lyer illusion by looking backwards and for-

[1] Experimental method described on p. 5.

wards along the figure (see Fig. 8), the eye movements were short and restricted over the under-estimated part of the figure, and long and more accurate over the over-estimated part. There seemed to be a difficulty in moving to the vertex of the acute angle in the middle of the line. This he attributed to the fact that the 'centre of attention' lies within the under-estimated part of the horizontal line, somewhere about the point P, and not at the point where the angular lines meet it. Or, to use R. MacDougall's term (96), the 'reflex centre of attraction', at which the tendencies to move either to the right or to the left are in equilibrium, may be said to lie within the under-estimated part. If this be the case, it must be deduced that the eye movements are regulated primarily by retinal sensations. The competing reflex attractions set up by peripheral sensations from the whole figure determine the fixation position of the eye; and the retinal sensations from the parts of the horizontal line situated on either side of the fixation position are unequal. Hence the illusion. Again with the Poggendorff illusion (see Fig. 8) (23), the first interrupting

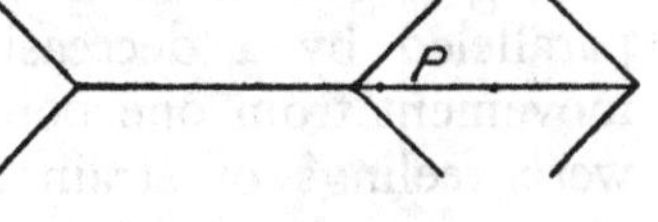

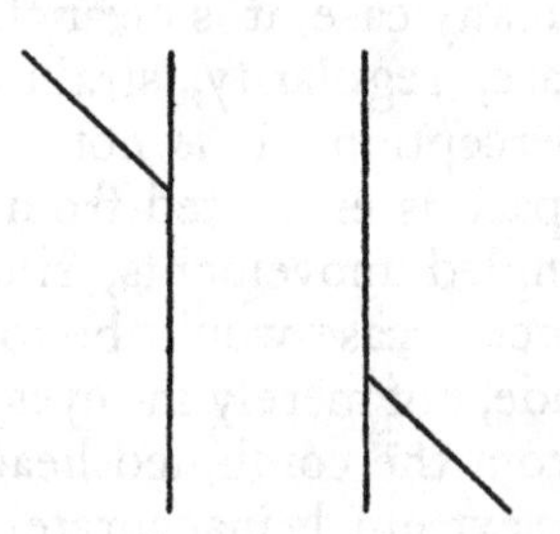

Fig. 8. Visual illusory figures.

vertical line produced a series of pauses and adjustments followed in general by a deviation of movement from the true direction of the oblique line. A reflex attraction is presumed to be set up by the radiating vertical line at its junction with the oblique line; and the subsequent direction of movement of the eye is determined as it were by the resultant of the forces set up by this reflex attraction and by the tendency to continue moving in the original direction. Again it seems to be the retinal sensations which initiate and control the movement, although the actual movements do affect subsequent perceptions. It may be deduced that the spatial arrangement of our perceptions is conditioned by the eye movements made as a reflex response to peripheral retinal stimulation. Actual kinaesthetic sensations of the changes of tension and strain produced by movement are not involved, and are probably not even present in consciousness. We are aware, even without complete recognition, of the retinal sensations preceding and following movement, but not of the movements themselves. Even the reflex response need not be accurately adjusted at first, since the impulse is merely to obtain clear vision of the object seen peripherally, not to bring its image to the centre of the fovea. Hence the first perceptions following these reflex eye movements are likely to be confused and to give inaccurate estimations of form and extensity, except with very familiar objects; and accurate estimate is only obtained after a large number of viewing movements. In support of this theory one may cite that familiar experience of casting one's eyes backwards and forwards from one end of an object to another when one wishes to judge its length. In all these judgments previous experience must of course play a very large part. We have come to know by experience that such a peripheral visual sensation followed by such and such a foveal perception corresponds to a definite spatial arrangement or judgment of extensity, by comparing over and over again these visual sensations with direct tactile sensations, until they are so well co-ordinated as to produce a unitary and accurate percept. Thus it was found by Grindley (62) that when the eye was momentarily stimulated peripherally, the position of the stimulus was always estimated to be more central and less peripheral, by a considerable angle, than was actually the case. This occurred not only when the eye remained fixated centrally during the estimation of position, but also when the eye moved outwards to the estimated position. The inaccuracy in the latter case must be explained by the fact that a peripheral stimulus normally remains in approximately the same position until the eye can move to fixate it; thus the original estimate of position is checked by the subsequent movement sensations and retinal impressions. In the absence of this check, no experience could be obtained in the correction of the estimation of the position. Hence the persistence of the error, which resulted in the first case from the tendency, shown in the last chapter to be very potent, to under-estimate the distance from the central position of a stimulus object in the periphery.

In a similar way we may be aided in judging the number of objects in a group by the number of eye movements necessary to include all the objects. Thus R. MacDougall(97) found that the estimated number of black squares in a group decreased if an orange square was placed in the centre of the group; he deduced that the eye tended to come to rest on the orange square instead of moving about among the black squares. The presence of two orange squares had the opposite effect; the conflicting peripheral sensations from them presumably stimulated reflex movement about the field. Movement also seemed to be facilitated by the arrangement of the squares in rows, particularly in horizontal rows; and such arrangements were accompanied by an over-estimation of the number of squares. Doubtless estimation of number is possible in the absence of any eye movement; but it seems to be clear from these observations that this estimation may be considerably influenced by eye movement.

(3) Perception of Depth and Distance

Our perceptions of depth and distance (away from us) are also partly dependent upon eye movement, in this case upon the movements of convergence and divergence; these are, however, only contributors to a number of co-ordinated factors. The most important sensory factor in binocular perception of depth is 'cross-disparity', the difference between the retinal sensations produced at the corresponding points of the two eyes, which have a slightly different view-point from one another. Cross-disparity alone, however, is not reliable; and it is probably assisted by the sensation of convergent adjustment in the binocular perception of depth and distance of comparatively near objects. Change of accommodation does not normally play a very important part, as may be proved from the inaccuracy of the estimation of depth or distance of an object when one eye is closed. If, however, the sight of one eye is destroyed, it seems probable that the individual learns to employ change of accommodation to a great degree in judgments of distance. With far distant objects, of course, none of these factors is of any utility, since the eyes are parallel and unaccommodated, and the disparate retinal images practically identical. In such cases the parallaxes obtained by moving the head are useful, but they are very largely interpreted in the light of experience. Judgments of great distances are notoriously deceptive when experience is at fault, as in an unusually clear or misty atmosphere.

But again it is the final resultant of the convergent adjustments, not the sensations of the individual movements producing them, which must contribute to depth perception. As with simple reactive movements, the final adjustment was found by Judd(79) to be produced by a number of preliminary inaccurate and corrective movements, made in response to peripheral retinal stimulation. These movements are not only unconscious; but individually they play no part in the resultant depth perception, which presumably is mediated by the changes of retinal sensation between the convergent and divergent positions, relatively different in the two eyes. These changes must be repeated several times to give accurate depth or distance perception, unless strongly reinforced by experience and familiarity.

(4) Perception of the Rate of Movement

Perception of movement is quite independent of eye movement, and appears to be a primary function of rod vision; but estimation of rate of movement seems normally to be dependent upon eye movement, and according to Dodge(38) is a function of the displacement of the moving object from the fovea. Dodge found that pursuit movements were set up in watching a moving object, but that these were interrupted at intervals by short, rapid, reactive movements, which occurred presumably when the image of the object had passed from the fovea to the periphery of the retina. One may surmise that the frequency and extent of these retinal stimulations has so frequently been related in experience, either to direct tactile sensation of the moving object, or to information from some outside source as to the rate of its movement, or to both, that at length the visual data give rise directly to a judgment of velocity. In a similar way we appear to be able to estimate our own rate of movement with relation to fixed bodies. The eye pursues a fixed object as long as possible; during this time retinal stimulation is foveal, and no estimation of speed of movement takes place, except such as is produced by the changes in tension of the head and neck muscles. Finally the image of the object passes into peripheral vision; the eye is jerked rapidly backwards to the primary position, and a new object is fixated. The frequency of alternation of these peripheral and foveal stimulations has presumably become associated with a priori judgments of, or information about, the speed of movement at which we are travelling (e.g. from the speedometer of a motor-car), until we are able to deduce the latter directly from this frequency.

Chapter IV

EYE MOVEMENTS IN READING

(1) General Description

We have already stated that the alternate movements and fixation pauses made by the eyes in reading are not dissimilar from the simple reactive type of movement and fixation. They are of course more regular, both as regards number and duration of the fixation pauses, and direction of movement. In general the eyes begin at the left-hand end of the line of print, pass from one fixation to the next along the line to the right-hand end, and then move smoothly back in a single sweep to the left-hand end of the next line. For a variety of reasons which will be described in due course, there is not always a regular succession of fixation pauses from left to right. The eye sometimes moves backwards in the contrary direction and refixates some part already covered of that line or of a previous line. Presumably these earlier words exert a stronger reflex attraction upon the eyes than do the succeeding ones.

The location and relative duration of the fixation pauses in the printed line are in part functions of fixed habits of movement, and tend to remain the same whatever the words in the line; but they are also affected by the nature of the reading material. Dearborn (32)[1] found that in a regular type of reading there was usually one long fixation near the beginning of the

[1] Experimental method described on p. 8.

line, though frequently not on the first word. This was followed by two or three short ones, and a longer one towards the end of the line. He concluded that the first fixation involved a general survey of the printed line; hence the immediately succeeding ones could be quite short. They might even last less than 100σ, since as Dodge(41) pointed out, the 'clearing up' process would have begun during the first long fixation of the line. Huey(72)[1] found that the first fixation was slightly within the line, sometimes on the second or third word; the last one was some distance from the end. The total indentation was usually about 18 per cent. of the length of the line; but the distribution of the pauses in the line was very irregular. With regard to the actual words fixated, Dearborn(32) found that the exact points fixated might be in any part of the words, or even in the spaces between them. They were merely significant as affording a centre about which were grouped the words perceived. Familiar words, and even phrases, required only a single fixation. Short connective and non-substantive words, prepositional phrases and relative clauses required the most or the longest fixations. Any words which stood by themselves and could not be fused into a language unit, such as a phrase, demanded separate fixations. Schmidt(128)[2] found that there was a tendency to fixate the apperceptive unit centrally. When the fixation point lay in a word space, the words surrounding it were grouped into a unitary percept. Certain words, such as prepositions, tended to enter into combination

[1] Experimental method described on p. 4.
[2] Experimental method described on p. 9.

with others. Thus we see that although there may be some conflict of opinion as to their exact locality, it is exceptional for fixation pauses to occur upon every word of the line. Though, as the result of ocular motor habits, the fixation pauses may tend to occur predominantly in certain positions, yet their actual location is finally determined by the apperceptive units into which the text is divided.

The average duration of the fixation pauses varies enormously according to a very large number of objective and more or less subjective factors. The averages given by various experimenters for several individuals reading a variety of types of material lie round about 300σ, which is, of course, very much greater than the average duration of the eye movements. This was shown above to be from 30 to 50σ, according to the length of the movement; it was constant for any one individual, in a constant state of health, and did not vary much from one person to another. There has been some difference of opinion as to whether the number of fixation pauses in a line is inversely proportional to their duration. Dearborn (32) stated that it was so. But Buswell (20) found that with children of sixteen or seventeen the correlation between number and duration of pauses was only $-0{\cdot}08 \pm 0{\cdot}055$, although with younger children it was $0{\cdot}49 \pm 0{\cdot}056$, that is, number and duration were inversely proportional. On the other hand, the writer (149) found that some adult readers made a rather large and variable number of short fixation pauses of regular duration, while others made a small and less variable number of longer pauses of irregular duration. The conclusion probably is that with readers such as Buswell's, whose maturity in reading varied greatly, there is no correlation between length and duration, because the more mature readers made fewer and shorter pauses than the less mature. But readers who are equally mature and practised in reading may exhibit a tendency towards either few and long pauses, or many and short pauses. [There is, however, no such relation between number and duration in the reading of a single individual.] The origin of the tendency will be discussed later. It is a relatively constant individual peculiarity.

A further measure of the variability in efficiency of the reading performance is afforded by the number of backward or regressive movements. In an average reading performance these seem to number about one per line, or one in two lines, but sometimes they may not occur for several lines together. The number varies a great deal. Four causes of these regressive movements are given by C. T. Gray(59) and Buswell(20):

(1) Failure to understand the general meaning of the content, or of particular words.

(2) Close scrutiny of the reading material if it is being studied in order to be remembered later, or to be analysed and paraphrased.

(3) Over-reaching and moving too far in rapid reading.

(4) Failure of motor habits, for instance, failure to move right back to the beginning of the line.

In the normal reading of adults the first three categories appear to be the most important; these can all be classed as showing some difficulty or conflict in the assimilation of the meaning of the content. In addition it seems probable that emotional reactions either to the content itself, or in whole or part to preformed moods or emotional dispositions, may affect the reading process directly. But this will be discussed more fully subsequently.

(2) Variation of Eye Movements with Objective Factors in Reading

The chief objective factor which affects the motor efficiency of the reading performance is the length of the printed line. The values in Table III are given by

Table III.

Experimentalists	Length of printed line in cm.	No. of pauses per line	
		Range	Mean
Erdmann and Dodge	8·3	3 to 5	—
	12·2	5 to 7	—
Huey	5·2	3 to 4	3·6
	6·0	3 to 4	3·6
	9·8	4 to 6	4·5
Dearborn	5·7	3 to 7	4·8
	9·6	5 to 8·5	6·7

Erdmann and Dodge (46),[1] Huey (72) and Dearborn (32) for various lengths of printed line. We see that on the whole there is an increase in the number of fixation pauses in the longer printed lines, but this increase is not appreciable unless the difference in length is about 4 cm. Since the majority of printed lines are

[1] Experimental method that of direct observation through a telescope.

from 9 to 11 cm. in length, the variation in practice is probably not large; it might average about 0·5 fixations per line. In any case, it does not appear that there is any increase in the number of fixations relative to the number of words in the line.

It seems reasonable to suppose that the number of pauses would be increased if the printing type were very small, simply because more words, and hence more reading material, would be crowded into each line. It was found by Gilliland(57)[1] that if the same number of words were printed in different sizes of type, readers with normal eyesight and normal eye movements showed little variation between 6 and 36 point[2] type. Thus evidently within wide limits the number of words, that is, the amount of material to be assimilated, was of more importance than their size or spacing. With extreme changes of type face, as in the substitution of Gothic face and handwriting for a Roman face, the number of fixations was shown by Gilliland(56) to increase by one or two per line, and the number of regressions to become excessive. It appeared that ocular motor habits were greatly deranged. This is only to be expected with such illegible type faces. But even less unfamiliar faces might give rise to some increase in the number and duration of the pauses because they would necessitate a closer scrutiny of the words.

[1] Experimental method approximately the same as that of C. T. Gray (see p. 10).

[2] Size of type is measured by the height of the tall letters in points. One point = 0·014 in. approximately; 10 or 12 point are the usual type sizes.

(3) Variation of Eye Movements with Age and Maturity of Reader

Of much more importance to the efficiency of the reading processes are the factors connected with the individual differences and peculiarities of the readers. It is convenient to consider first the effects of age, or rather of age as it affects maturity and practice in reading. The figures in Table IV are taken from the results of the work of Schmidt(128), C. T. Gray(59),[1] and Buswell(20)[2]. The exact ages are a little doubtful, since all the figures are given for United States School Grades; but they tally for the three experimentalists. It appears that very little if any decrease in the average number of fixation pauses takes place after the age of nine or ten years, and not much decrease in the average duration. The decrease in the average number of regressions, however, is continuous until fifteen years of age, and there is some indication that it may continue to the adult stage. But on reaching this age, the variations are probably due to factors other than maturity and practice in reading. Interesting records are given by Buswell(20) of the reading of the youngest children. It appears that even at this age some children possess the rudimentary habits of adult reading. Fixations are longer and more numerous, but they occur quite regularly in order; usually there is one to each word, but sometimes a group of two or three words will require only one fixation. At the other extreme are those children

[1] Experimental method described on p. 10.

[2] Experimental method approximately the same as C. T. Gray's.

Table IV.

Age of reader	Average number of pauses per line as found by			Average duration of pauses in secs. as found by			Average number of regressions per line as found by	
	Schmidt	C. T. Gray	Buswell	Schmidt	C. T. Gray	Buswell	C. T. Gray	Buswell
Elementary School								
6			18·6			0·660		5·1
7			15·5 }14·9			0·432 }0·485		4·0 }3·8
8			10·7			0·364		2·3
9		10·0	8·9		0·284	0·316	2·4	1·8
10	}6·3	9·1 }8·9	7·3 }7·4	}0·314	0·250 }0·266	0·268 }0·262	2·1 }2·0	1·4 }1·5
11		10·0	6·9		0·276	0·252	2·4	1·3
12		7·5	7·3		0·250	0·236	1·4	1·6
13		7·8	6·8		0·272	0·240	1·5	1·5
High School								
14			7·2			0·244		1·0
15	}7·0	}6·4	5·8 }6·2	}0·311	}0·230	0·248 }0·241	}0·8	0·7 }0·8
16			5·5			0·224		0·7
17			6·4			0·248		0·7
College								
18 and over	6·5	6·9	5·9	0·308	0·226	0·252	1·1	0·5

who can scarcely build the separate letters into words. They make twenty, thirty, sometimes forty or more fixations in a single line of eight or nine words; and their eyes, instead of fixating each letter or word regularly in turn, oscillate wildly backwards and forwards over the line with no apparent plan of action.

There is some indication that the increase of efficiency of the motor processes is due to the increasing effects of practice, in addition to the maturing of

Table V.

Number of reader	Number of pauses per line		Number of regressions per line	
	Before training	After training	Before training	After training
1	11·7	7·8	2·4	1·0
2	7·1	5·1	1·1	1·1
3	9·5	7·2	1·0	0·4
4	6·4	6·0	1·6	1·1
5	5·0	3·5	0·3	0·2
6	8·4	6·5	2·5	1·6

ability with age. Thus O'Brien (106) gave children aged about ten to twelve years intensive practice in rapid, silent reading, and also training in short exposure work, in which a gradually increasing amount of material was read at short exposures. The duration of the fixation pauses seems to have decreased little, if at all, but the figures in Table V show that the number of pauses and regressions decreased considerably. As might be expected, the worst readers showed the greatest improvement—but not sufficient to make them as good as the good readers. O'Brien states that there was a marked increase of the speed of normal reading, with slight increase in comprehension of the material read, and of ability to answer questions on it afterwards. We do not know, however, whether this speed of reading and decrease of number of fixations and regressions were retained permanently, together with an adequate comprehension of the material read; and if that were so, whether it was at the cost of an unduly increased mental strain or physiological strain on the eyes.

(4) Differences of Eye Movements in Silent and Oral Reading

As might be expected, the differences between oral and silent reading are large. The values given in Table VI are taken from the results of Schmidt(128) and Buswell(20). It will be seen that after the earliest stages of reading the average number and duration of the pauses and the average number of regressions are higher throughout for oral than for silent reading. During the earliest stages, almost all practice in reading is oral. Hence the child when reading silently articulates habitually in the same way as in reading aloud, and the ocular motor processes are thus identical in the two cases. After this age, oral reading is found to be much slower than silent reading; and, according to Pintner(120), less is remembered of what is read orally. In oral reading, no correlation is found between rate and comprehension, as is found in silent reading. The rate of reading seems to be determined by the rate of articulation in oral reading, and by the

Table VI.

Type of reader	Average no. of pauses per line as found by				Average no. of regressions per line as found by	
	Schmidt		Buswell		Buswell	
	Oral reading	Silent reading	Oral reading	Silent reading	Oral reading	Silent reading
Elementary School:						
6 to 8	—	—	14·2	14·9	3·3	3·8
9 to 13	8·1	6·3	9·4	7·4	1·7	1·5
High School	8·6	7·0	8·7	6·2	1·4	0·8
Adult	8·2	6·5	8·4	5·9	1·2	0·5

Type of reader	Average duration of pauses in secs. as found by			
	Schmidt		Buswell	
	Oral reading	Silent reading	Oral reading	Silent reading
Elementary School:				
6 to 8	—	—	0·557	0·485
9 to 13	0·398	0·314	0·314	0·262
High School	0·372	0·311	0·268	0·241
Adult	0·382	0·308	0·300	0·252

rate of assimilation in silent reading. Judd (81) found that there was generally one pause per word in oral reading, with a long pause at the end of the line for the voice to catch up the eyes. This result supports the work of Quantz (123)[1] on the eye-voice span. Quantz (123)

[1] The eye-voice span is equivalent to the number of words over which the voice lags behind the eyes. Quantz measured its extent by slipping a card over the reader's page at a point determined by the experimenter, and counting the number of words which were spoken when the view was cut off.

found that this was narrowest at the end of the line of print, so that presumably the eyes had paused for the voice to catch up. Buswell(19), however, found that the span varied with the sentence rather than the line unit; that is, it was widest at the beginning of the sentence and narrowest at the end. Probably both types of variation can occur. Buswell(19) also found that a wide span was characteristic of good and rapid oral reading, and a narrow one of poor and slow oral reading. One may explain this in the following way. With the good reader the rate of assimilation is high, hence the rate of reading, apart from vocalization, is rapid, and the eyes are able to advance at a good pace along the printed line. The rate of vocalization is limited, however, by the necessities of articulation and cannot exceed a certain rather moderate value.

Thus there is a considerable interval of time between perception and vocalization. Assimilation takes place shortly after perception, and time is still available for deducing the correct pronunciation of the words from the assimilated meaning of the context. If the eye-voice span is very narrow, so that articulation takes place before the meaning has been fully assimilated, mispronunciation may occur, particularly of words which can be pronounced in different ways, such as 'lead', or 'read' in the present and past tenses. It is also likely that words will be wrongly emphasized, and the balance of the sentence upset. The effect of this mispronunciation is naturally to set up regressive movements, since it will be necessary to look back to find out the correct pronunciation. Hence the excess of regressions in oral reading over silent reading, although the rate of the former is so slow.

We see then that the motor processes are not only much slower in oral reading, but they may lack the regularity of those in silent reading, owing to the long pauses necessary for the voice to catch up the eyes, and the regressions resulting from mispronunciations, etc. This would be of comparatively little importance to the adult reader, who is rarely called upon to read orally, were it not that the tendency to articulate persists in adult reading in the form of 'inner speech'. This inner speech is usually very much more rapid and slurred than normal speech, and is by no means necessarily accompanied by lip movement. There has been much controversy as to whether it is universally present in adult reading, and whether it can be eliminated with practice. Curtis(28) recorded the movements of the larynx during silent reading by means of a tambour pressed against it, and found that movements of the vocal chords habitually accompanied reading. Secor(129) found that silent reading was not affected by simultaneous whistling or recital of the alphabet, which he assumed would eliminate inner speech; but it is not certain if this would be so. Pintner(121) found that if numbers were recited aloud, silent reading was at first much disturbed, but with practice became as fast and accurate as when inner speech was undisturbed. When ordinary reading was recommenced, some articulation reappeared, but it was less than before. It does not appear that much was gained by this artificial elimination of inner speech. There seems to be little doubt that inner speech of some kind always occurs, frequently without any great retardation of the reading; but in the best readers it may be to some extent replaced by auditory motor imagery of the words and sentences. On the other hand, strongly articulated speech, accompanied by lip movement, such as is found in immature readers, or readers of a pronounced motor type, generally produces a slow rate of normal reading. As in oral reading, the rate of reading is limited by the mechanical necessities of articulation, rather than by the rate of assimilation. Schmidt(128) was of opinion that the differences in the number and duration of the fixation pauses in silent and oral reading would have been much greater, had it not been for the persistence of strongly marked inner speech. It follows from this that training in oral reading should not take the place of training in silent reading at any time, and should not be continued for too long, lest it produce the prominent and retarding type of inner speech.

(5) Variation of Eye Movements with the Purpose of the Reading

If we confine ourselves to silent reading, the type of ocular motor process seems to vary greatly with the purpose of the reading. Ordinary rapid reading is different from scanning, careful reading and detailed studying, summarizing, paraphrasing, analysing for style, etc., and proof-reading. No actual study has been made of the movements of the eyes in scanning or skimming. Whipple and Curtis(154) showed that the rate of skimming was highly correlated with the normal rate of reading. But different individuals adopted very different devices, even though they were all mature and highly educated readers. Thus some individuals generally read about half of each sentence, and inferred the rest. Occasionally whole sentences might be omitted if they appeared unimportant; for instance, one reader omitted all quotations, and particular instances of general laws, while another omitted well-known details. Other readers read vertically down the centre or edge of the column, selecting words that had some distinctive form, or appeared to be important, and pausing on them. One reader omitted arbitrarily; he chose a size of 'jump' which he thought would fit the material, and employed this throughout, reading only the words which came at the end of a 'jump'. It was usual to fixate any artificially distinguished material, such as italicized words, words surrounded by quotation marks (except by the subject mentioned above), or followed by an exclamation mark, or preceded by a capital letter (as in proper names). Clearly the eye movements would differ considerably for these different readers. The number and duration of the fixation pauses would probably be greater the more 'condensed' the content of the reading matter.

C. T. Gray(59) found that the number and duration of the fixation pauses increased with one subject, as shown in Table VII, when he was required to reproduce afterwards the general thought of a piece of prose. The number of fixations and regressions increased still more when he was required to answer questions afterwards, but the increase in the duration of the pauses was not so great, showing that there was a tendency to read in smaller units when required to answer questions than when required to reproduce the thought.

Judd and Buswell(82)[1] found that their readers, ranging in age from about twelve to eighteen, reacted in several different ways when told to read a paragraph carefully, so that they could answer questions about it afterwards. Examples of their procedure are given in Table VIII. Some readers, e.g. 1 and 2, changed their procedure very little from that of ordinary reading. Some, e.g. 3 and 4, increased both number and duration of fixation pauses, others increased either length (7 and 8) or number (5 and 6). It was, however, more usual to increase number and decrease duration of the pauses than vice versa; that is, the reading units became smaller. The number of regressions increased in most cases (2, 3, 4, 5 and 6). It appears that these readers had little idea of the correct procedure to adopt when they wished to read

Table VII.

Purpose of reading	Average number of pauses per line	Average duration of pauses in secs.	Average number of regressions per line
Normal reading	8·1	0·266	1·8
Reproduction	8·7	0·292	1·8
Answering questions	9·6	0·278	2·4

[1] Experimental method approximately the same as that of C. T. Gray (see p. 10).

carefully. Quite probably many of them merely felt a kind of restless anxiety in face of the difficulty of remembering. Others apparently tried to memorize verbally parts of the reading material—hence the small reading units and the number of regressions. Judd and Buswell state that these children had never been taught the proper procedure to adopt in such circumstances. Consequently we can only deduce

Table VIII.

Number of reader	Average number of pauses per line in		Average duration of pauses in secs. in		Average number of regressions per line in	
	Rapid reading	Careful reading	Rapid reading	Careful reading	Rapid reading	Careful reading
1	10·6	10·5	0·224	0·220	2·0	1·9
2	7·5	7·8	0·304	0·308	0·7	1·4
3	6·6	9·1	0·272	0·316	0·5	0·9
4	8·8	14·0	0·224	0·240	0·3	2·5
5	6·9	11·5	0·248	0·232	0·9	2·8
6	4·3	6·1	0·244	0·252	0·4	1·1
7	10·4	10·0	0·224	0·256	2·6	2·1
8	9·0	7·6	0·232	0·252	2·1	1·8

that in their case the wish to read carefully resulted in a decrease of efficiency of the ocular motor processes.

This decrease is small, however, compared with that which occurred when the readers were required to paraphrase what they had read. It appeared that nothing which could properly be called reading occurred at all. The number of fixation pauses increased to from twenty-seven to sixty-four per line, usually with some increase in their duration, and regressions were very numerous. Apparently the eyes wandered aimlessly backwards and forwards over the printed line while the reader thought what he should say. Some lines showed much more irregularity than others. A rather similar mode of behaviour was adopted when the reader was required to analyse the reading material for purposes of grammar, or in order to study the style and vocabulary, although the confusion, particularly in the latter case, was not nearly so great. In fact it appeared that if the reader was instructed to analyse in any way or to look for something in the printed page, other than the general meaning and significance of the content, there was a tendency for this random oscillation to occur, indicating the existence of what Judd and Buswell termed a 'period of confusion'. It seemed to show a complete return to the procedure of the immature and unpractised reader; and it was accompanied by a kind of general nervous commotion, or emotional turmoil, which appeared not only in the eye movements but also in increased muscular agitation of the head. Judd and Buswell attribute these phenomena to the lack of training of the readers in the particular type of analysis required, since some of them were much more proficient in certain types than others. But it seems possible that the prospect of a difficult and exacting task may first have set up unfavourable emotional reactions which in turn affected the objective performance, breaking down the acquired motor habits of the mature reader, and producing a regression to the childish type of behaviour. This seems a more probable cause for so inefficient a type of behaviour than mere lack of familiarity.

A rather similar type of behaviour, involving considerable irregularity of reading and many regressions, was found by the writer (150) to occur when a reader encountered unexpectedly a number of misprints in the reading text. He regressed to find out if he had really seen a misprint, or to try and discover what the correct word should have been. But the experienced proof-reader was much less affected. Normally his reading was very regular; and it only became slightly more irregular when misprints were encountered. Neither was the number of regressions greatly increased. He was able to react in a systematic and effective manner, without becoming wildly confused, and without altogether losing the gist of what he was reading. It seemed that the presence of misprints made him more alert and watchful with regard to the typographical details of the text, without making his eye movements confused and irregular; and that he was able, with long practice, to attend primarily to these details, while assimilation of the general meaning of the content occupied only the fringe of his consciousness. In this the experienced proof-reader differed from the inexperienced, whose behaviour was confused and irregular because he did not know the correct method of dealing with material containing misprints, and was unable to concentrate upon the typographical details without struggle and conflict.

(6) Variation of Eye Movements with the Nature of the Reading Material

(*a*) *The Native Language.* It is natural to expect from the results previously discussed that if a reader finds a passage difficult to read, his motor performance will be to some extent affected. Clearly ease or difficulty is not a quality which can be absolutely determined. There are a number of factors which contribute to the ease or difficulty of a passage, and

Table IX.

Paragraph number	Average number of pauses per line made by reader			Average duration of pauses in secs. made by reader		
	1	2	3	1	2	3
1	5·2	7·0	6·0	0·216	0·204	0·232
2	6·2	8·2	4·2	0·220	0·220	0·296
3	5·8	11·2	6·3	0·256	0·220	0·268
4	5·6	6·2	7·4	0·240	0·260	0·272
5	6·8	11·0	8·2	0·296	0·276	0·316

their effects upon different individuals may not be the same. Thus Judd and Buswell (83) obtained the results given in Table IX for the silent reading, by three readers aged between thirteen and seventeen, of a series of paragraphs which are said to have been graded experimentally, from 1 to 5, for difficulty in oral reading. It appears that although on the whole there was a steady rise from paragraph 1 to paragraph 5 in the duration of the fixation pauses, the increase in number of fixations was irregular. Moreover, the reactions of the three readers were different from each other. But as we do not know what constituted the ease or difficulty of these paragraphs to these readers—no introspections being recorded—we can draw no deductions.

In a further series of observations by Judd and Buswell (82), four readers of about the same age as the above read a number of different types of material, with no particular instructions as to how to read them. It is clear from the values given in Table X[1] and from Fig. 9[1] that certain types of material affected the reading processes for some readers, others for others. If we study the average total fixation times per line (obtained by multiplying together the average number of fixations per line and their average duration), we observe that the values for Reader 3 were fairly uniform throughout; so also was his average number of regressions. Thus he appeared to be capable of dealing adequately with all these types of reading material, although, as might be expected, he read fiction the most rapidly. Reader 4 read easy and blank verse more slowly and French grammar much more slowly than the other material. But normal reading habits do not seem to have been impaired, except perhaps by French grammar, for the number of regressions was not unduly increased. Reader 1 read blank verse, French grammar and algebra much more slowly than the other material, and made a large number of regressions. But he regressed even more in reading fiction. Apparently he reacted by making a number of very long fixations. Reader 2, on the other hand, made a very large number of short fixa-

[1] After the figures given by Judd and Buswell.

tions, and a very large number of regressions. He appears to have been the poorest reader generally; his procedure was inadequate in reading verse, French grammar and algebra, but particularly blank verse.

Table X.

Type of reading material	Average number of pauses per line made by reader					Average duration of pauses in secs. made by reader				
	1	2	3	4	Av.	1	2	3	4	Av.
Fiction	6·1	8·5	6·2	8·0	7·2	·228	·196	·224	·236	·221
Geography	7·3	11·2	7·9	8·5	8·7	·236	·208	·252	·252	·237
Rhetoric	8·6	11·7	7·7	8·3	9·1	·240	·200	·236	·236	·228
Easy verse	9·4	13·1	8·4	10·0	10·2	·244	·200	·268	·272	·246
Blank verse	11·9	16·8	8·5	9·6	11·7	·272	·208	·260	·252	·248
French grammar	10·6	14·1	8·0	11·8	11·1	·300	·204	·280	·272	·264
Algebra	12·5	14·4	9·5	8·1	11·1	·264	·212	·244	·264	·246
Average	9·5	12·8	8·0	9·2	9·9 M.V. 1·31	·255	·204	·252	·255	·241 M.V. ·011
	Av. 9·9 M.V. 1·48					Av. ·241 M.V. ·019				

Type of reading material	Average number of regressions per line made by reader					Total fixation time per line in secs. made by reader				
	1	2	3	4	Av.	1	2	3	4	Av.
Fiction	3·0	2·1	0·2	0·4	1·4	1·39	1·67	1·39	1·89	1·59
Geography	1·1	3·2	0·65	1·2	1·5	1·72	2·33	1·99	2·14	2·05
Rhetoric	1·5	3·7	0·6	1·5	1·8	2·06	2·34	1·82	1·96	2·05
Easy verse	1·9	3·7	0·3	1·2	1·8	2·29	2·62	2·25	2·72	2·47
Blank verse	2·6	6·1	0·5	1·4	2·7	3·24	3·49	2·21	2·42	2·84
French grammar	2·3	4·0	0·4	2·3	2·3	3·18	2·88	2·24	3·21	2·88
Algebra	3·1	4·5	0·6	1·0	2·3	3·30	3·05	2·32	2·14	2·70
Average	2·2	3·9	0·5	1·3	2·0 M.V. 0·4	2·45	2·63	2·03	2·35	2·37 M.V. 0·40
	Av. 2·0 M.V. 1·1					Av. 2·37 M.V. 0·18				

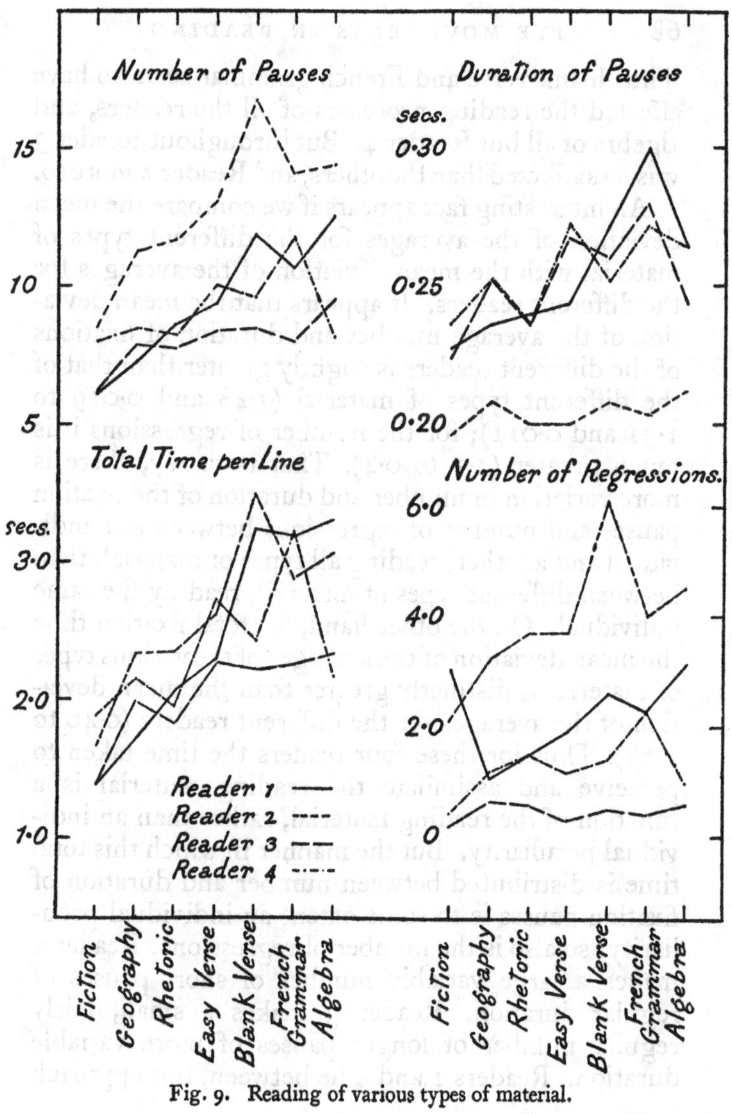

Fig. 9. Reading of various types of material.

Thus blank verse and French grammar seem to have affected the reading processes of all the readers, and algebra of all but Reader 4. But throughout Reader 3 was less affected than the others, and Reader 2 more so.

An interesting fact appears if we compare the mean deviation of the averages for the different types of material with the mean deviation of the averages for the different readers. It appears that the mean deviation of the average number and duration of fixations of the different readers is slightly greater than that of the different types of material (1·48 and 0·019 to 1·31 and 0·011); for the number of regressions it is much greater (1·1 to 0·4). That is to say, there is more variation in number and duration of the fixation pauses and number of regressions between one individual and another, reading all types of material, than between different types of material, read by the same individual. On the other hand, for total fixation time the mean deviation of the averages of the various types of material is distinctly greater than the mean deviation of the averages for the different readers (0·40 to 0·18). Thus for these four readers the time taken to perceive and assimilate the reading material is a function of the reading material, rather than an individual peculiarity. But the manner in which this total time is distributed between number and duration of fixation pauses is to some extent an individual peculiarity; so also is the number of regressions. Reader 2 makes a large variable number of short pauses of regular duration. Reader 3 makes a small, fairly regular number of longer pauses of more variable duration. Readers 1 and 4 lie between, but approach more nearly to Reader 3 than to Reader 2. It also appears that on the whole Reader 2's method is less efficient than Reader 3's.

It would have been interesting to know whether these variations between different types of material were due to differences in comprehension of the material, or of familiarity with it, or of interest in it. Judd and Buswell unfortunately did not obtain any introspections on the subject. They appear to have ascribed these differences unquestioningly to a lack of training in reading certain types of material; the reader who was interested in a certain subject would have received more training and have had more practice in reading on that subject. But this seems to assume altogether too much. It is quite possible that Reader 4 may have hastily skipped through the algebra he disliked, whether he understood it or not. Again, how many of those sufficiently familiar with the grammar of the French language to read about it with tolerable fluency would derive any interest from this pursuit? And may not Reader 2 have greatly enjoyed reading blank verse, and lingered over it in order to appreciate thoroughly its sound and rhythm, or the thoughts and imagery which it aroused? In support of this possibility may be mentioned the case of a reader who did the writer's experiments on eye movements in reading (149), and who highly appreciated the style and rhythm of certain prose passages. In the reading of one of them it appeared at first sight that the reading times were large and variable, and the number of regressions considerable. But it was found that these reading times varied quite regularly in three-line periods, not as a result of sentence divisions, which were not coincident, but simply in response to the rhythm and swing of the prose.

An attempt to distinguish between the effects of comprehensibility and interest was made in this work of the writer's on eye movements in the reading of highly educated adults (149).[1] The readers were given a number of different pieces of English prose to read, and were required to report their introspections afterwards. It was found that the reading performance varied considerably according as to whether there was any effort to understand the material. In Fig. 10 are shown the times taken to read successive lines of print (reading time per line), and also the number of regressions in these lines. (*a*) and (*b*) show the reading of passages which the readers made little effort to understand or take in, because the topic did not interest them. (*c*) and (*d*) show the reading of passages where comprehension was difficult owing to lack of context leading up to them, but where there was considerable effort to understand. It appears that in (*a*) and (*b*) the reading time did not vary greatly, but gradually decreased throughout the reading—that is, the rate of reading increased; and there were few regressions. In (*c*) and (*d*) the reading time varied greatly from line to line, and there were a large number of regressions.

In Fig. 11 is shown the reading of two passages, the first (*a*) with understanding but no interest, the second (*b*) with both understanding and critical interest. (*a*) is seen to resemble (*a*) and (*b*) in Fig. 10,

[1] Experimental method described on p. 12.

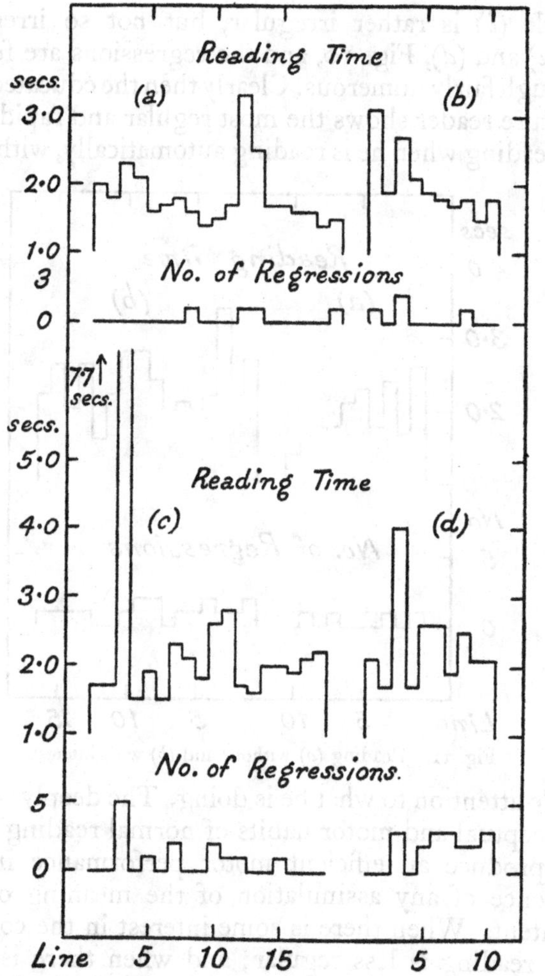

Fig. 10. Reading (*a*), (*b*), without comprehension, and (*c*), (*d*), with comprehension difficult.

while (*b*) is rather irregular, but not so irregular as (*c*) and (*d*), Fig. 10, and the regressions are fewer, though fairly numerous. Clearly then the educated and mature reader shows the most regular and rapid type of reading when he is reading automatically, with very

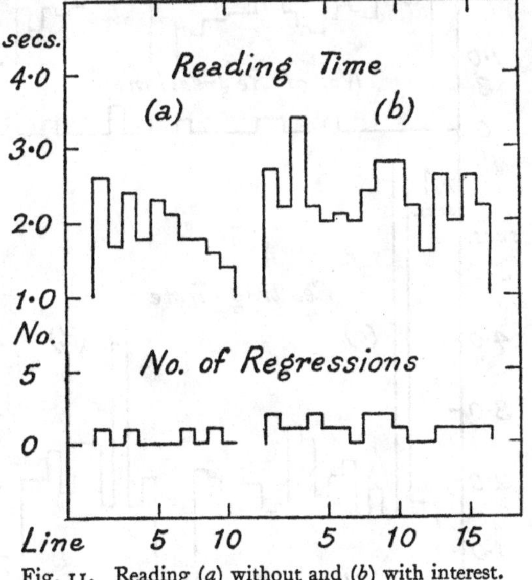

Fig. 11. Reading (*a*) without and (*b*) with interest.

little attention to what he is doing. The deeply rooted perceptual and motor habits of normal reading seem to produce an efficient motor performance in the absence of any assimilation of the meaning of the content. When there is some interest in the content the reading is less regular; and when there is considerable effort to understand, it becomes markedly irregular.

The interest in a piece of reading material was not always dependent upon the actual meaning and significance of its content. Quite frequently the latter

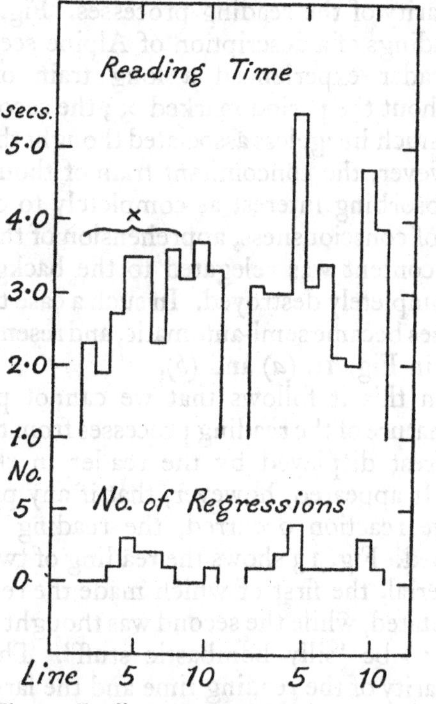

Fig. 12. Reading accompanied by imagery and associated thought.

aroused more or less irrelevant trains of thought and imagery; but these were of great personal interest to the reader, and seemed to be allied to some emotionally tinged body of cognitive material already present in the mind. If these trains of thought existed in consciousness together with an attempt at apprehension of the meaning of the content, a conflict usually ensued, which was accompanied by much irregularity of the reading processes. Fig. 12 shows two readings of a description of Alpine scenery. The first reader experienced a long train of imagery throughout the period marked ×; the second experienced much imageless associated thought throughout. If, however, the concomitant train of thought was of such absorbing interest as completely to occupy the centre of consciousness, apprehension of the meaning of the content was relegated to the background, or even completely destroyed. In such a case the reading processes became semi-automatic, and resembled those shown in Fig. 10 (*a*) and (*b*).

From this it follows that we cannot predict the actual nature of the reading processes from the amount of interest displayed by the reader in the general topic. It appeared, however, that if any pronounced affective reaction occurred, the reading was much disturbed. Fig. 13 shows the reading of two passages of material, the first of which made the reader angry and irritated, while the second was thought by another reader to be 'silly bombastic stuff'. The extreme irregularity of the reading time and the large number of regressions are apparent.

It seems then that reading is least regular when some type of struggle is going on—whether in the effort to understand, or as a result of a conflict of interests, or in connection with some strong affective reaction. Similarly it was shown by Judd and Buswell(82) that the most irregular types of children's reading occurred in the difficult task of analysing and paraphrasing, when much effort and conflict were apparent. This may perhaps indicate that the source of

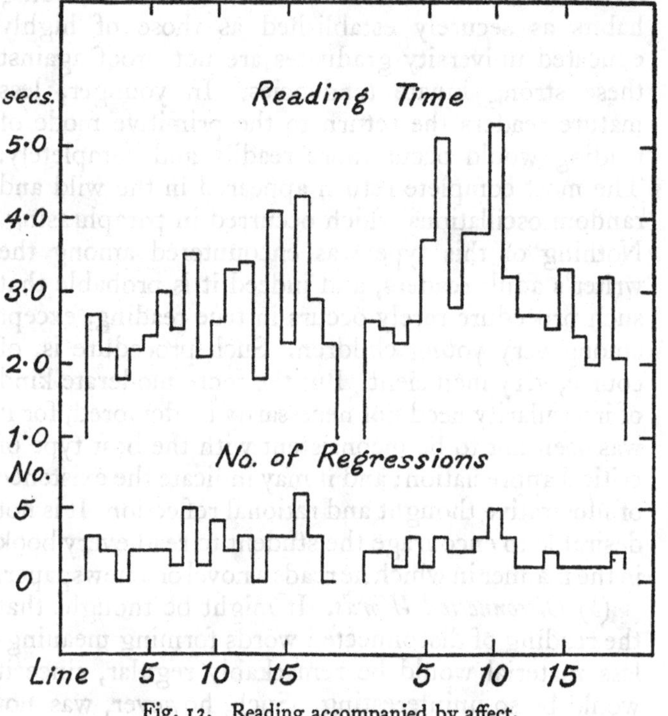

Fig. 13. Reading accompanied by affect.

these irregularities is some type of powerful conative impulse, the disturbance being more pronounced the greater the strength of the impulse. The imperfections in the reading processes would be the result of innate deficiencies rather than of lack of training; or rather, these innate impulses would dislocate the acquired motor and perceptual habits of efficient reading, and cause the reader to regress to an earlier and more immature mode of behaviour. It seems that reading habits as securely established as those of highly educated university graduates are not proof against these strong innate tendencies. In younger, less mature readers the return to the primitive mode of reading would occur more readily and completely. The most complete return appeared in the wild and random oscillations which occurred in paraphrasing. Nothing of this type was encountered among the writer's adult readers, and indeed it is probable that such procedure rarely occurs in true reading, except among very young children. Such procedure is, of course, very inefficient. But the more moderate kind of irregularity need not necessarily be deplored, for it was seen not to be inconsistent with the best type of critical appreciation; and it may indicate the existence of illustrative thought and rational reflection. It is not desirable to encourage the student to read every book in the manner in which he reads a novel or a newspaper.

(*b*) *Disconnected Words.* It might be thought that the reading of disconnected words forming meaningless material would be remarkably regular, since it would be so uninteresting. Such, however, was not by any means always the case. Fig. 14 (*a*) shows the reading of disconnected meaningless material, by one of the adult educated readers mentioned above (149), to have been slow and irregular. The readers were required to read every word, but frequently there were more pauses, on the average, than there were words, and their duration was greater than in normal reading. Apparently normal reading habits were so disorganized that the reading time, unregulated by these habits, became long and variable. In general

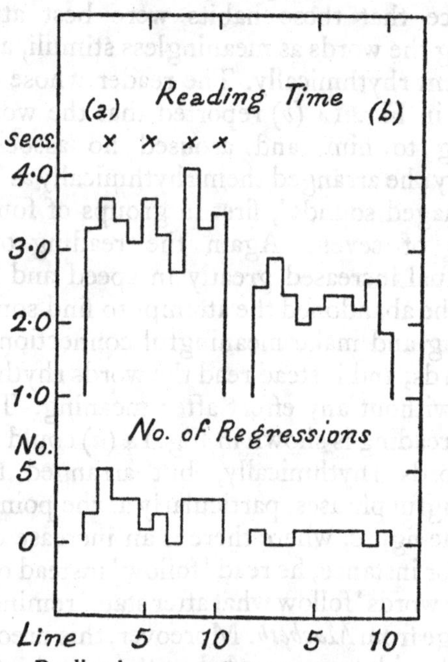

Fig. 14. Reading by two readers, (*a*), (*b*), of disconnected meaningless material.

the regressions were few, and when they did occur they were only short, covering one word or part of a word, and could be distinguished from the longer regressions typical of normal reading. Some readers, however, as shown in Fig. 14 (*b*), were fairly rapid and regular throughout in their reading of disconnected material, while others became more rapid with practice. Apparently then it was possible to develop certain temporary habits of reading meaningless material. It seems possible from the introspective evidence that these habits were best attained by treating the words as meaningless stimuli, and grouping them rhythmically. The reader whose reading is shown in Fig. 14 (*b*) reported that the words meant nothing to him, and aroused no associations or imagery; he arranged them rhythmically as 'meaningless imaged sounds', first in groups of four, then in groups of seven. Again the reading of another individual increased greatly in speed and regularity when she abandoned the attempt to find some general meaning and make meaningful connections between the words, and instead read the words rhythmically in pairs, without any effort after meaning. The reader whose reading is shown in Fig. 14 (*a*) could not group the words rhythmically, but arranged them into meaningful phrases, particularly at the points marked × in the figure, where there is an increase of reading time; for instance, he read 'follow' instead of 'fellow', and the words 'follow what after state' reminded him of a passage from *Macbeth*. Moreover, this preoccupation with the ideas aroused by the words frequently seemed to set up a struggle which suggested the presence of some strong conative tendency, similar to that encountered in the reading of material not clearly understood.

It appears then that a series of eye movements can be executed regularly either by arranging the words according to the sentence pattern, or according to some deliberately adopted method of rhythmical grouping. It is fairly clear, however, that different sets of habits are utilized in the two cases, since the fixation pauses were always longer and more numerous in the reading of disconnected material than of connected material. Thus when connected meaningful material is read with no apparent attention to the meaning of the content, the words are not treated as isolated stimuli. But the habits of movement are so strong that, once the process has begun, movement can continue perfectly regularly and automatically.

(*c*) *Foreign Languages.* The movements of the eyes in reading foreign languages seem to resemble those made in reading rather incomprehensible English prose. The figures in Table XI are given by Judd and Buswell(82) for the silent reading, by six readers aged about sixteen, of passages of easy French prose and of prose in which French and English phrases alternated (mixed prose). It appears that the reading of French prose was fairly similar to the reading of mixed French and English prose; and the number and duration of the fixations were of the same order as the number and duration in the reading of French grammar (see Table X). Indeed, the duration of the fixations did not seem to vary greatly; and the number of fixations in both cases varied from eight or nine to fourteen or fifteen per line. The reading was tolerably straightforward; although regressions were rather numerous, they were not excessive. It appeared that these individuals could read French with fair fluency, though rather more slowly than difficult English prose.

Buswell, in a later study (22), found that if younger children were taught French they had great difficulty in attaining such comparatively regular reading habits. If they employed their normal English reading habits they were unable to understand what they read; consequently the numbers of fixations and regressions were greatly increased. If, however, they were taught

Table XI.

Number of reader	Average number of pauses per line in reading				Average duration of pauses in secs. in reading			
	Mixed prose		French prose		Mixed prose		French prose	
	1st passage	2nd passage	1st passage	2nd passage	1st passage	2nd passage	1st passage	2nd passage
1	10·2	10·7	12·2	11·1	0·272	0·256	0·272	0·288
2	12·1	13·1	14·4	10·2	0·240	0·284	0·260	0·264
3	12·9	15·1	13·6	11·6	0·284	0·272	0·264	0·284
4	12·7	13·1	14·1	12·4	0·252	0·256	0·244	0·260
5	11·5	14·7	11·7	10·1	0·220	0·252	0·272	0·268
6	9·2	10·4	9·7	10·2	0·240	0·236	0·248	0·232

French by the 'direct method', they were able to develop normal reading habits much more readily than if taught by the 'indirect method'.

Figures are also given by Judd and Buswell (82) for one of the above readers reading Latin prose, and mixed Latin and English (see Table XII). These figures are stated to be fairly typical of the other readers in the class. It is clear that a different type of procedure existed here. The number of fixations is comparable to that encountered in paraphrasing. Regressions were very numerous, and the eyes oscillated wildly backwards and forwards over the line, as in the periods of confusion in paraphrasing. Moreover, head movements were numerous, showing that nervous tension was high. It is clear that the readers made little attempt to read; they merely tried to pick out familiar words, to analyse the situation into its elements, and invent some kind of connected meaning from their jumble of impressions. That even this analysis was unsuccessful appears from the wildness of the oscillation and wandering of the movements. Buswell (22) found later that this type of procedure was largely a product of learning by the 'indirect method'. The reader became completely dependent upon the 'vocabulary' at the end of the book for looking up unfamiliar words when translating, and was quite at a loss how to proceed without it. If, however, he was taught by the 'direct method', he could learn to read Latin normally though rather slowly after two years' training. It should be noted, however, that since the construction of the Latin sentence is so different from that of the English—and French—sentence, frequent regression is unavoidable until the Latin language is so familiar that the reader is able to *think* the meaning

Table XII.

Type of material	Average number of pauses per line	Average duration of pauses in secs.
Mixed prose, 1st passage	16·7	0·232
Mixed prose, 2nd passage	39·1	0·276
Latin prose	45·4	0·288

according to that particular construction and arrangement of words and phrases.

(*d*) *Spelling*. A study has recently been published by Abernethy(1)[1] of the movements of the eyes in learning to spell difficult words. She selected good and poor spellers of about twelve years of age, and required them to study certain words, which they had misspelt in a previous test, until they felt they could

Table XIII.

Type of reader	Average number of pauses per word	Average duration of pauses in secs.	Average percentage of regressive movements	Average learning time in secs.	Average percentage of correct letters	
					Before learning	After learning
Adults	24·8	0·352	11·0	8·812	80·3	93·6
Children						
Good spellers	28·1	0·548	25·4	15·432	85·5	95·6
Poor spellers	44·8	0·492	23·5	22·020	76·1	77·6

spell them correctly. Some adults were also required to study difficult scientific words, with which they had not previously been acquainted, until they could spell them correctly. She obtained the figures shown in Table XIII for number and duration of fixations and number of regressions made while studying the words; the average time taken to learn the words; and the average percentage of letters given correctly before and after learning. The first observation which strikes one is that the poor spellers did not learn to spell the words at all. This somewhat vitiates the results. It

1 Experimental method the same as that of C. T. Gray (see p. 10).

seems that the poor spellers must have possessed some deeply rooted habit of misspelling, or some emotional or other disability which prevented them from spelling correctly. The adult subjects studied the words more straightforwardly, with fewer regressions, than did the children. When the location of the fixations was examined, it was found that the adults and the good spellers among the children recognized their errors, and made a systematic attack on the difficult parts of the words; they also made frequent surveys of the word as a whole. The poor spellers wandered vaguely through the word without any particular plan of attack. Some spellers studied mainly by syllables, others by letters; the adults tended to adopt the broader unit, and study by syllables. These observations give another instance of the aimless nature of the poor student's procedure. Clearly these individuals not only could not spell, but had no idea how to learn to do so by making a strong effort to remove the source of error.

(*e*) *Mathematics.* It was shown by Dearborn(32) that the average number of digits covered at each fixation pause is less than the average number of letters. Indeed, the work of Cattell(24)[1] would lead us to expect that, except with familiar number sequences such as 1000, or 1066, which could be treated as units, the number of digits perceived at a single fixation would be about the same as the number of words. Terry(141)[2] has confirmed this. He showed

[1] See below, p. 111.

[2] Experimental method approximately the same as that of C. T. Gray (see p. 10).

that in general one, two or three digits were read at each fixation pause, though one exceptional reader could read as many as four. It followed that in reading groups of digits placed in horizontal lines the number of the fixation pauses was very much greater than in normal reading. Moreover, the reading pauses were in general much longer. Terry also demonstrated the occurrence of short 'guiding' pauses, used probably for locating the first and last digits of the numeral. They were particularly frequent

Table XIV.

	Number of digits in numeral						
	1	2	3	4	5	6	7
Average number of pauses per numeral	1·15	1·2	1·9	2·4	2·9	3·7	4·15
Average duration of pauses in secs.	0·386	0·458	0·540	0·526	0·568	0·538	0·516

with long numerals; and also occurred almost invariably at the beginning of the line. In general the number of reading pauses per numeral increased with the length of the numeral; the average duration also increased up to three-digit numerals, but then remained stationary. In Table XIV are shown the average values for five readers. The increase in pause duration varied for the different readers; some continued to increase steadily throughout, or increased and decreased irregularly. Terry found that the quality of familiarity, as in the numerals 1000 and 25,000, reduced the average number of pauses, and also to some extent pause duration. It was much more effective in this respect than mere regularity of grouping, as in the numerals 99, 33 and 637,637. It was found that individual peculiarities manifested themselves in the number and duration of the fixation pauses. From Table XV it appears that some individuals used many short fixations, and others few long fixations; the former also made a relatively large number of guiding pauses. Judging from the total reading time, the method of frequent short pauses seemed to be the

Table XV.

Type of reader	Number of reader	Total reading time in secs.	Total number of pauses	Average duration of pauses in secs.
Many short pauses	1	30·98	76	0·408
	2	32·84	85	0·386
Few long pauses	3	34·56	51	0·668
	4	35·94	69	0·521
	5	38·74	67	0·578

more efficient, although more regressions were made by these readers than by those who made few and long pauses.

Terry also studied the movements of the eyes in reading and solving arithmetical problems each containing two numerals of varying numbers of digits. He found that in general the problems were first read fairly straight through; the characters of the numerals were noted, and the general relationship of their digits, but these digits were not accurately perceived and assimilated. This method of procedure he termed a 'partial first reading'. But sometimes the actual digits were recognized and remembered fully and in detail; these cases were termed 'whole first readings'. A 'whole first reading' generally necessitated a larger number of longer pauses than a 'partial first reading'. In fact the former approximated to the type of procedure which occurred in the reading of lines of numerals, where the conditions were such that the reading was of the quality of 'whole reading'. The

Table XVI.

Number of digits in numeral	Average number of pauses per numeral		Average duration of pauses in secs.		Average time in secs. required to read individual numerals	
	In problem	Isolated	In problem	Isolated	In problem	Isolated
1	1·00	1·15	0·233	0·386	0·233	0·429
2	1·17	1·20	0·314	0·457	0·370	0·574
3	1·50	1·90	0·329	0·541	0·493	0·961
4	1·83	2·40	0·329	0·526	0·603	1·193
5	1·66	2·90	0·308	0·569	0·513	1·555
6	3·08	3·70	0·229	0·538	0·703	1·881
7	3·83	4·15	0·293	0·516	1·123	2·091

accelerating effect of 'partial first reading' is shown by comparing the number and duration of the fixations, and the average time required for reading the numerals, in problems and in lines of isolated numerals (see Table XVI). The average time required to read the numerals in problems is only a little over half that required to read the isolated numerals. It is true that interest in reading and solving the problems may have had a considerable effect in accelerating the reading of these numerals, but probably this effect manifested itself in the large number of 'partial first readings' in problem solving. It was found that in general the longer numerals received a slightly greater number of 'partial readings' than 'whole readings'; but that the shorter one and two-digit numerals were given 'whole first readings', since they necessitated only a single fixation as a rule. Some readers reported far more 'partial first readings' than others, and the former required a shorter time to read the numerals, and made fewer fixations, than the latter. In general, the former read the words of the problems, and also ordinary prose, more rapidly than did the latter. 'Partial first reading' is probably a rapid method of dealing with arithmetical problems which is acquired at a comparatively late stage; the method of 'whole first reading' is more primitive, but persists in the less efficient individuals.

In addition to the whole or partial first reading, re-reading was frequent, both for the verification of details and for copying prior to computation. Some individuals habitually re-read the numerals, but others proceeded to compute from the context of the problem, whether the first reading had been partial or whole. That is to say, the habit of 'partial first reading' was not necessarily connected with the habit of direct computation, although presumably the most rapid workers would make use of both. The individuals in these experiments were not always consistent, however; they sometimes varied their methods. It appeared that during computation one numeral of the two was made a 'base of operations', and was fixated more frequently and for a greater length of time than the other; presumably during this period the arithmetical processes were performed.

Thus it appears that the highly skilled computer, just as the highly skilled reader, has succeeded in evolving a rapid and accurate series of motor habits which are well adjusted to the functions which the eye must perform. The habits are not the same in the two cases, but are equally suitable to the processes of perception and assimilation which they subserve. It is clear that the efficient individual attends to the general meaning, in the first case of the reading content, in the second, of the problem. It was found by Wilson (156a) that children frequently would not trouble to read properly the wording of an arithmetical problem, but only attended to the figures. This was because the wording was dull and obscure. The problems should be presented as concrete situations likely to interest the child; just as in teaching the child to read, the reading content should be of sufficient interest to the child to encourage him to read for its meaning.

Buswell (21) has made a study of the movements of the eyes in adding columns of figures. Here again movements were much shorter and fixations much longer than in normal reading; sometimes their duration was as great as 4 secs. But in addition by good arithmeticians the fixations were relatively shorter and much less variable than in addition by poor arithmeticians. The former made about one fixation per digit, and at times grouped two digits together in a single fixation; while the latter made a large number of fixations whenever they encountered any difficulty. Difficult number combinations required longer fixations, also; probably as much from lack of practice in using them as from inherent difficulty. With the good computers the fixations progressed regularly up and down the column of figures starting from somewhat below the top (or the bottom) of the column; there were few regressions, and scarcely any 'periods of confusion'. The poor computers made regressions both at the beginning of the columns and within them, and sometimes a series of irregular and random movements up and down the column, similar to those encountered in the 'periods of confusion' which occurred in analysis and paraphrasing, and in the reading of Latin.

Tinker(142)[1] found that individuals reading lines containing algebraic formulae made 4·5 more fixations per line, and nearly three times as many regressions, as in reading lines of algebra which did not contain formulae. The increase in duration of the pauses was only moderate, however, and did not vary much between different individuals. The number of fixations showed great individual variation. Some readers, although skilled in dealing with algebraic formulae, showed a tendency towards analytical reading, and the breakdown of normal reading habits, as exhibited by Judd and Buswell's readers with Latin. The reading of scientific prose containing formulae was, in the case of mature students, very similar to the above. But it was found that readers with only a small knowledge of chemistry frequently just glanced at these formulae without really assimilating them, or

[1] Experimental method the same as that of Miles and Shen (see p. 11).

even skipped them altogether. Consequently there was a much smaller excess of number of fixations and regressions over those in normal reading for these readers than for the adult students with a considerable knowledge of chemistry. Thus clearly it is not always advisable for the reader to employ the regular and rapid ocular motor habits of normal reading, or he will omit the most important parts of his task.

In reading lines of isolated algebraic formulae, it was found that normal reading habits were completely abandoned. As in the case of Buswell's readers adding columns of figures (21), the duration of the fixation pauses, as well as the number, was greatly increased to as much as 2 to 4 secs., and was also very variable. This large increase in pause duration shows that a different mode of procedure was adopted from that employed in reading prose containing formulae. It was necessary to fixate almost every letter and digit of the formulae, paying particular attention to fractions and exponents. The eye movements were thus much shorter than usual, and the whole line was covered from beginning to end, not omitting the first and last few letters as in normal reading. In fact, the formulae were read more as a design is studied. The fixations were, however, irregular in number and distribution, and the number of regressions considerable. Probably more regular and efficient habits would be developed if reading of lines of isolated formulae were more frequently practised. It has been shown that whenever a type of reading has become efficient through use and practice, an adequate mode of motor behaviour has been acquired. This appeared in spelling, in reading numerals and formulae (in text), and in adding columns of figures. The last case is particularly instructive, because eye movement in a vertical direction is comparatively infrequent, and is generally much less regular and accurate than horizontal movement. Yet it appeared that when this movement was employed as part of a habitual act of reading, it was as accurate as the horizontal movement of normal reading. Thus again, if the whole task of perception and assimilation and the resulting processes of thought was accurately and efficiency discharged, the ocular motor processes were also well adapted to the end and object of the activity.

(7) Variation of Eye Movements with Motor Habits

It was mentioned that the location and relative duration of the fixation pauses in the printed line were partly regulated, in normal reading, by fixed habits of eye movement. Dearborn(32) considered that after reading had continued for a time 'short-lived motor habits' were established, which regulated to some extent the number, position and duration of the fixations, subject of course to the variations produced by apperception and assimilation, which have already been described. These motor habits were also responsible for regulating the length of the return movement from the end of one line of print to the beginning of the next. Thus where the lines were of unequal length or irregularly indented inaccurate return movements were liable to occur, because 'short-lived motor habits' could not readily be formed. It was found by the writer (149) that there was a greater tendency at the beginning than at the end of the reading of a piece of material for the return movements to be either too long or too short. Dearborn (32) also stated that this inaccuracy increased if the lines of print were unusually long; this seems probable in view of the excessive inaccuracy of wide-angled voluntary movements (see p. 19). Huey (72) found that fast readers in particular tended to form 'short-lived motor habits', and read several pages of a book at almost identical rates. Dearborn considered that the rate and regularity of reading were not necessarily correlated, but that on the whole the faster readers made longer pauses at the beginning of the line and few pauses subsequently. Huey was of opinion that these habits of reading could be greatly speeded up by voluntary effort, i.e. that either the number or duration of the fixation pauses could be greatly decreased. It seems probable that such a procedure would, in the mature and habituated reader, be very difficult to maintain without either much loss of comprehension or undue fatigue.

The importance of permanent motor habits in reading was shown by comparing the reading of connected prose and disconnected words (see p. 77). It appeared that, even in the absence of any real apprehension of the connected material, perception was sufficient to mediate regular and efficient motor reading habits.

There seems to be some indication that different permanent motor habits may have been acquired by different individuals. It was found by the writer (149) that readers could be roughly divided into two groups,[1] each reading at about the same average rate; the first tended to make a small regular number of long fixations of irregular duration, and the second to make a large and variable number of short fixations of fairly constant duration. On the whole, the readers of the

Table XVII.

	Average reading time per line	Average standard deviation	Average number of pauses per line	Average standard deviation
Group 1	2·28	0·52	7·7	1·5
Group 2	2·34	0·52	9·5	2·2

	Average duration of pauses in secs.	Average standard deviation	Average number of regressions per line	
			Nonsense	Sense
Group 1	0·29	0·043	0·4	0·7
Group 2	0·25	0·035	0·8	1·2

first group made fewer regressions than those of the second; but this was not always the case. The averages for the two groups are given in Table XVII. The results of Judd and Buswell, described on p. 68, were rather similar. One of their readers made a large variable number of short regular fixations and a large number of regressions. One made a small regular number of long irregular fixations, with very few regressions. The other two readers were intermediate, but approached the second of the above

[1] Each containing five persons.

cases. Again, in reading lines of numerals some of Terry's readers made many short pauses, and others made few long pauses (see p. 85). Tinker's readers, however, all tended to vary their number of pauses while keeping the duration relatively constant (see p. 89). But on the whole it appears that it is a permanent and established individual peculiarity to read either with a large number of short pauses, or a small number of long pauses. In normal reading the second

Table XVIII.

	Average ranking in				
	Average number and duration of pauses and their standard deviations	Average number of regressions	Steadiness of voluntary fixation	Accuracy of voluntary movement	Accuracy of return movement
Group 1	3·5	4·2	3·0	2·8	3·8
Group 2	7·4	6·1	6·5	6·9	6·3

method appears to be more efficient, on the whole; but in the reading of numerals the first is superior. This peculiarity is not merely determined by practice and maturity in reading, since in the writer's experiments both groups of readers contained intelligent and highly educated adults. There was some indication that these peculiarities were functions of the motor efficiency of the ocular muscles. In Table XVIII are given the average ranks of the readers of the two groups mentioned above in average number and duration of fixations and their standard deviations, and average number of regressions. (In number of fixations and regressions the readers were ranked from the smallest number upwards, but in duration they were ranked from the largest duration downwards, since number and duration are more or less inversely proportional.) In addition are given the average rank in steadiness of voluntary fixation on a point, in accuracy of voluntary movement from point to point, and in accuracy of return movement in reading from the end of one line of print to the beginning of the next. Thus the readers of the second group, who made a large number of short fixations, were in general less steady than those of the first group in voluntary fixation, and less accurate in voluntary movement and in the return movements from the end of one line to the beginning of the next. It is quite natural that those who are unable to maintain steady fixation and to make accurate movements should make the shortest possible movements and fixations in reading. This inaccuracy and unsteadiness were not directly connected with any known ocular defects of the readers. They might have been produced by some more subtle ocular disability which would have been revealed by a very thorough ophthalmological examination of all the readers. Or it is possible that there are innate differences of efficiency in the motor activities of the ocular muscles, similar to the innate differences of skill in the activities of the muscles of the hand and arm.

The tendency to regress did not seem necessarily to be connected with this lack of efficiency of the ocular muscles. Among the readers of Judd and Buswell and of Terry, regressions were made rather more frequently by the readers who made a large number of short pauses than by the others. But among the readers doing the writer's experiments, one reader in Group 1 and two in Group 2 showed a habitual tendency to overrun the word by a short distance and then regress to it; this appeared as much in the reading of disconnected words as in the reading of connected prose. The two readers in Group 2, one other reader in Group 2, and one in Group 1, showed a tendency to make longer regressions when confused by the meaning of the connected prose. Thus the habitual tendency to overrun the word seems not to be necessarily related either to the number and duration of the fixations, or to the tendency to make longer regressions when confused by the meaning. It may of course be due to faulty motor habits acquired when learning to read, but no data are available on this subject. One cannot dismiss the possibility that the employment of a large number of short fixations is due also to the method of learning to read. But it does seem significant that it is correlated with inaccuracy and unsteadiness of voluntary movement and fixation, in which no habitual motor procedure has been developed. But whatever the origin, it seems clear that the use of frequent short pauses, or less frequent long pauses, and the tendency to overrun the word and then regress to it, are permanent ocular motor habits, unconnected with perception and assimilation of the reading content.

Chapter V

VISUAL PERCEPTION IN READING

(1) Visual Perceptual Processes

Before passing to a study of visual perception in reading, it is desirable to consider briefly visual perception in general. Perception has provided a subject of great controversy since the earliest times of psychological speculation. It is not proposed to relate to the reading process all the diverse theories put forward in the controversy, but to confine this survey to a short résumé of the results of some of the experimental work which has been done upon the process of perceiving visually. The major part of this experimental work has been carried out with the tachistoscope, an instrument varied in design, but intended to expose to the observer the object to be perceived instantaneously and for only a very short interval of time. Under normal conditions, the perceptual process is so rapid, complex and uncontrolled that it is extremely difficult to analyse. With a very rapid exposure, the process may not be carried to its ultimate conclusion, the complete recognition and assimilation of the object exposed. Thus the resulting partial perception may throw some light upon the intermediate stages in the process; or else the stages are sufficiently retarded by the difficulty of perception to be open to introspection. Tachistoscopic perception is, however, subject to the criticism that can be applied to so much of the simplification introduced into psychological experimentation—namely, that the parts or stages which appear under its unfamiliar and artificial conditions cannot be assumed to exist in the familiar and rapid process of normal perception. Thus the results which are now to be described must in all probability be qualified considerably before they can be applied to the normal process of visual perception.

Various experimenters have deduced, from the introspections of their subjects upon the process of visual tachistoscopic perception of conventional designs or drawings of real objects, the existence of a number of stages which progress and develop into one another. It is probable that in the perception at least of unfamiliar objects the majority of these stages can be distinguished at one time or another. But the order of their occurrence is not always fixed or definite. Indeed sometimes two or more stages seem to alternate in occupying the centre of consciousness, or one may occupy the centre while the other is felt to exist simultaneously at the fringe. The object to be perceived is experienced first as a visual pattern; it exists in space, and has a general extent and position, a kind of flat clearness. This might be termed the stage of 'something there'. Next it is recognized as an object, and given an 'objective reference'; parts of the field stand out slightly but significantly, and become differentiated from the rest. This is termed by Dickinson (36) and Freeman (50) the stage of the 'generic object'; but since the term 'generic' has been used with a different meaning in connection with imaging, it seems preferable to employ here the term 'objective reference'. After this, the details become highly specific; they are recognized as appertaining to some one object in particular. This is called the stage of the 'specific object'. Finally the object is itself recognized and named; there may be a feeling of 'I know what you are' before the actual name is given. This recognition develops from the recognition of the logical meaning of the salient parts, which rise more and more clearly out of the field, while the remainder fades into the background (Dickinson(36)). With a very familiar object the first stages are much telescoped, and recognition is almost immediate. It seems probable that the 'objective reference' stage is omitted altogether, but the naming stage increases in prominence. It is clear that the process by which the object to be perceived rises out of the general field of vision is of much importance, but it is difficult to explain its nature.

In addition to these primary constituents of the perceptual process, Rogers(124) found that there occurred with some frequency during the later stages two important types of secondary cognitive process, namely kinaesthetic and organic sensations, and imaginal processes. The kinaesthetic sensations were either of a general, diffuse bodily type; or specifically localized, the verbal and ocular being the most frequent. Organic sensations were comparatively infrequent in occurrence, and their distribution was general. Kinaesthetic sensations occurred in the great majority of perceptions; their chief function was held by Rogers to be 'to bear effort or intent to find-significance-in-the-figure, or to question the fitness of meaning ascribed by other processes'. But with some observers they were wholly or partially replaced by auditory or tactual imagery. It is to be inferred that the general kinaesthetic and organic sensations were a function of the strong conative impulses which usually exist during tachistoscopic perception, and are to a great extent a product of the experimental situation. The more familiar the percept and the more habitual the perceptual situation, the fainter and less essential would be these kinaesthetic and organic sensations. On the other hand, in so far as the search for meaning was carried on verbally and was accompanied by naming, so far would kinaesthetic verbal sensations or auditory verbal imagery necessarily be present. Again, Rogers found that, with a complex stimulus object, interpretation and naming were usually followed by appreciation or valuation of the perceptual experience. Such appreciation very frequently occurs during perception, although less sophisticated observers often describe it in terms of the object perceived. It gives rise to affective and conative processes, which are generally accompanied by kinaesthetic and organic sensations.

Rogers found that even with very simple visual stimuli, imagery, especially visual, occurred in almost every perception, although comparatively late in the course of events. The function of the imagery was to interpret and elaborate the meaning and significance of the percept. With more complex and meaningful stimuli, the processes of interpretation and orientation of the percept in accordance with past experience increased in importance and frequency. Hence, also, imagery became richer and more varied. It has, however, been stated by other observers that they rarely, if ever, experience visual imagery, and that verbal imagery is not essential; in which case the perceptual process seems to be attended and elaborated by some species of imageless thought. But it is probable that most individuals employ a considerable amount of visualization and verbalization.

Judd (80) considers that even the simplest form of perception contains certain tendencies to reaction; and these motor tendencies are often of the greatest importance. Past perception has been followed by a series of bodily reactions, and these in turn have given rise to tendencies to similar reaction as a result of perception in the present. As was shown in Chap. III, the perception of space seems to be conditioned by the reflex tendency to bring the optical image of the object to the fovea, since reflex movements of this kind have been associated in the past with this type of perception. Thus it appears that kinaesthetic imagery can indeed be inherent in the process of perceiving visually.

It was found by Bartlett(7) that when perception was only partial, it was described as a 'feeling of' or an 'impression of' some figure or object. This seems to indicate that perception had only proceeded as far as the 'specific object' stage. In addition to the first stage of vague apprehension, there was usually a faint feeling of familiarity or ease of apprehension, involving some attribution of meaning. With the fuller development of the perception occurred a partial freeing of the content from the background of sensation. This might occur either through imaging, which related the percept to definite situations or objects not actually present; or through thinking, which indicated some apprehension of general and abstract relations. The former was thus more definite and concrete than the latter; moreover it was found to be accompanied by strongly marked affective tone, usually at a minimum during thinking. This affective tone seems to have developed from the 'feeling of familiarity' or ease of apprehension which qualified the vaguer forms of perception; it was the basis of criticism and valuation. The three levels of immediate sense experience, imaging and thinking, were normally all present during perception, one developing out of the other.

Perception appears always to involve some attribution of meaning, or reference of the present percept to the body of past experience. This reference is mediated by some association of similarity between the present and past experience. But clearly there are innumerable past experiences similar to any present percept. It was shown that the reference might either be to some concrete object or situation, or to some generalized system of ideas. Which species of reference is adopted varies from situation to situation, though it is in part a permanent individual characteristic. But the reference is to some extent determined by the nature of the stimulus; it seems that the more concrete the object presented, the more readily will it be referred back to some concrete experienced situation. And since such a reference occurs very frequently in normal life, there must be a strong tendency for association to take place in this direction, particularly with percepts of a familiar nature. Such association will probably be direct and rapid. Another factor of some importance is the cognitive clearness or comprehensibility of the percept. If there is some difficulty in understanding the meaning of the percept, there will probably be confusion in consciousness while the relevant experiences are recollected and compared. The various stages in the process of perception which were detailed above may take place successively and fully. Thus the more abstract processes of thinking will be multiplied; association will be broadened and complete assimilation delayed.

Although there is some difference between the various experimenters in their enumeration and description of the different stages of perception, there is a broad agreement as to the occurrence of the following processes:

(1) Perception of gradually increasing specificity, as the salient parts rise clearly out of the background.

(2) Kinaesthetic and organic sensation, either general or localized, or both.

(3) Imagery, predominantly visual, but sometimes also kinaesthetic or auditory-kinaesthetic.

(4) Associated imageless thought.

(5) Feeling of familiarity, ease of apprehension, affective tone.

(6) Attribution of meaning and complete assimilation.

The sixth stage might precede any of the others, except the first; the more familiar the stimulus object, the more speedily would this stage occur.

No treatment of the nature of perception can be adequate without some consideration of the theories advanced by the German school of Gestalt psychology, of which Koffka and Köhler are the chief exponents. It has long been known that the percept varied characteristically according to certain spatial relationships of its parts. Goldscheider and Müller(58) and others have shown that a greater number of objects can be perceived if they are grouped regularly, according to some familiar spatial relationship, such as that of symmetry, than if they are arranged irregularly. Thus there is always a tendency to group or arrange the diverse parts of a complex percept according to some such familiar general relationship or pattern. But the Gestalt psychologists[1] consider that patterns, structures or configurations (Gestalten) are primary qualities of the organization of sensation; and that organization in the sensory field originates as a characteristic achievement of the nervous system. It appears in the young child as the result of maturation of the nervous system, and is in no way the result of learning. The first forms to be perceived by the infant are those which are biologically significant to it, and this biological significance constitutes the primary meaning of the form or configuration. Secondary meanings are acquired whenever parts of the field change their aspect or significance because they are included in the activity of the moment (cf. Judd's

[1] This description of Gestalt psychology is compiled, it is hoped correctly, from the following sources: Koffka, K., *Psychol. Bull.* (1922) (84); Koffka, K., *The Growth of the Mind* (85), and Köhler, W., *Gestalt Psychology* (86).

views as to the importance of tendencies to reaction, described on p. 101 above).

The stability of these sensory organizations or configurations is shown by the fact that their form remains constant if spatial position, size, colour and brightness are changed, provided that they are not seen too far out in peripheral vision. Moreover, the actual size of the configuration will appear to remain constant while the retinal image is greatly increased or decreased by change of distance. The only thing which cannot be changed, if the same form or configuration is to be perceived, is the special set of relations of retinal stimulation which are decisive for the segregation of a definite whole. For the remembered forms of familiar objects, certain forms are always favoured; these are generally geometrically simple and physically outstanding, such as the right angle. Thus a table seen in perspective is always thought of as square; and a child or an unskilful adult will generally draw it as such. Although the total field of vision forms a single configuration or whole, it can nevertheless be analysed into genuine parts which themselves form wholes or groups which can again be analysed, and so on. In general, equal and similar objects tend to form unitary wholes separated from what is dissimilar to them. Also the wholes tend to be simple and regular, and to form closed areas. Moreover, there is a definite tendency for part of the field to emerge as the 'figure', while the remainder constitutes the 'ground' (Rubin (125)). In general, the figure lies somewhere near the centre of the ground. The figure appears to be more solid and vivid, and more organized and structuralized, than the ground. It usually forms a 'closed' figure—that is, it is complete in itself, and is enclosed by boundary lines. It has more of a 'thing' character, while the ground has a 'substance' character. The figure is always the more resistant, while the ground may fade away into a vague background. Figure and ground together form one of the primitive sensory organizations mentioned above. The actual constituent parts of figure and ground are to some extent interchangeable. By adopting a special attitude towards the total field, certain parts of it can be selected and kept together as the figure, while others are suppressed, and constitute the ground. But this sensory organization may also change without the exertion upon it of any outside influence; for processes which remain the same for some time may alter the conditions under which they occur and block their own path. Thus if we look at the accompanying diagram (see Fig. 15) we may, at will, see as the 'figure' either a Maltese cross, or a cross with slender diagonal arms. If, however, we continue to regard it without adopting any special attitude towards it, the two figures will alternate.

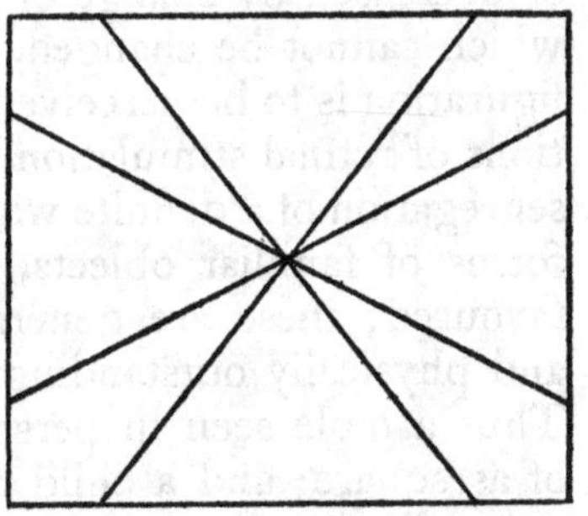

Fig. 15. Reversible 'figure' and 'ground'.

The functions usually attributed under experimental conditions to attention can be ascribed to such attitudes. The observer, on entering the experimental situation, has in readiness certain modes of response which are themselves 'structures' or 'configurations'. His attitude, which can be controlled largely either by suitable manipulation of the experimental situation or by suitable instructions, determines his readiness to carry out specific processes. Thus if he is instructed that the stimuli of a series will all be either greater or less than each other, very few judgments of equality will be given. But under normal conditions, when he is not called upon to make comparisons, judgments of equality will be numerous. Thus attention to any particular feature of the perceptual stimulus, such as equality or inequality, is equivalent to a readiness to perceive that feature according to a 'structure' previously determined by certain instructions or experimental conditions.

The influence upon the nature of the perception of the preconceived attitude and orientation of the individual's mind has been stressed by other psychologists, particularly by Bartlett(8). These attitudes may be generalized, affectively toned reactions to the conditions under which perception is taking place, whether these are experimental or the normal conditions of everyday life. Such are the attitudes of doubt, confidence, suggestibility, aggressiveness, and so on. They may be more specific, and largely cognitive in character, such as the attitudes just described. Or again, perception may be affected by some more deeply rooted reaction. If this is primarily affective, it will take the form of a mood, an enduring emotional 'set' of elation, depression, etc., which has been set up by conditions independent of the immediate perceptual situation. On the other hand, it may be primarily cognitive, and appertain to the large body of organized cognitive material which constitutes the interests, based upon a temperamental foundation. Instances without number can be adduced of the interpretation and valuation of percepts in accordance with some strong interest. Indeed it seems probable that it is they which ultimately determine the interpretation of all the primary sensations. Habituation, familiarity and comprehensibility may affect the rate of perception, and purely affective attitudes may influence the subsequent valuation. But the actual interpretation must result from some combination of cognitive, affective and conative processes, and the most individual, specific and permanent forms of such combinations are the interests, and their resultant organization, the character.

Clearly the conception of perception and sensation put forward by the Gestalt psychologists differs greatly from those described earlier in the chapter, although it appears that there may be more points of contact than Köhler(86) at least would allow. The chief conflict seems to lie in the source or origin of these configurations or organizations, the older schools holding that they are set up by experience upon the primary basis of unorganized sensation, while the Gestalt psychologists affirm that they are primary qualities of sensation, originating at the level of the nervous system. The typological school of Jaensch and his co-workers endeavours to show that both these rival theories are right up to a point. Jaensch(73) considers that the starting-point in the development of perception is formed by sensations that are permeated with higher mental processes. These primary perceptions, which he terms 'eidetic images', approach very close to true memory images, and are only differentiated completely from them at a comparatively late stage of development. These eidetic images are literally seen, as an after-image is seen, and they reproduce the original stimulus object with extraordinary vividness and clearness of detail; their occurrence and stability do not, however, depend upon optical factors, but upon subjective factors such as meaning, interest, attitudes, and so on. In certain pronounced cases there may be actual confusion of eidetic images with perceptions of real objects. In many individuals eidetic imagery is found in a more or less overt form until puberty, and may persist in the adult stage. These individuals Jaensch has termed 'integrates'; they are characterized by a co-operation and integration of all the subjective mental functions, and an 'intuitive' interpretation of the external world, in which the perceptions are dominated by the total circumstances of the situation, objective and subjective, rather than by a single perceptual stimulus. But the opposite type of individual, the 'disintegrate', has passed through this stage either in early infancy, or hereditarily, in which case he receives the structure of the objective perceptual world as a completed heritage. His perceptions are built up piecemeal from separate parts; they are not dominated by subjective thought processes, and do not take the form of eidetic images. Jaensch considers that this typological distinction is fundamental; all individuals belong to one of these two types, or to subdivisions of them, although their characteristics may be latent, particularly the eidetic characteristic of the integrate. It follows that the disintegrate will tend to explain the development of perception in terms of experience based upon pure primary sensations. The integrate, however, will think in terms of innate configurations or combinations of sensory and perceptual factors. Neither type can appreciate the outlook of the other, but each must learn to tolerate the other.

It is clear that the forms which are perceived in reading can in no way belong to the category of primary sensory organizations or configurations. But as the letters and words are learnt, the primitive sensations of meaningless groups of straight lines, curves and angles must be organized intentionally by the reader into definite groups and configurations which are each intimately related to some language form. The configurations must acquire the meanings of these language forms, and then become spontaneously organized together through association of meaning. Thus the process of perception in normal reading is somewhat different from that described at the beginning of the chapter. For the language forms, and the corresponding kinaesthetic sensations or imagery, must be implicit in the process, even if they are subliminal; whereas we saw that in other types of perception kinaesthesis was only a secondary process, and not essential to perception. In reading, the stages of perception are much telescoped, and recognition and assimilation are very rapid. Moreover, the processes of imagery and associated thought, of interpretation and valuation, which follow on the primary stages of perception, are very rich and varied, since they are taken over, together with the acquired meanings of the perceptual configurations, from the corresponding language configurations of word, phrase and sentence. The adult reader seems to pass directly from the visual perception to these meanings and processes of thought, but the intermediate language stage is probably always present in a rudimentary form; and in the young child it is prominent.

(2) The Nature of the Perceptual Processes in Reading

It is now necessary to consider what parts of the visual field we actually perceive when we read. It seems improbable that we perceive individually every letter of every word. Thus Cattell(24)[1] found that during a tachistoscopic exposure of 0·01 sec., the average reader could perceive three to four single letters or digits, two disconnected words containing up to twelve letters, or a sentence of four words. But one exceptional reader could perceive five to seven letters or digits, four disconnected words, and sentences of seven words. Erdmann and Dodge(46) obtained similar results with German readers, although using an

[1] The work of Cattell, Erdmann and Dodge, Goldscheider and Müller, Zeitler, and Messmer is discussed at great length by Huey in *The Psychology and Pedagogy of Reading*, Chaps. IV and V(72). For fuller treatment of this experimental work, the reader is referred to this source.

exposure of 0·1 sec. Thus three or four times as many letters could be reported when they were grouped into words as when they were disconnected. Indeed, Cattell found that it took longer to name single letters than single words; while Messmer (100) and Kutzner (89) found that the normal rate of reading is very similar to, or is less than, the rate of reading disconnected letters. But this, as Huey (72) points out, is principally because we are unaccustomed after the early stages of reading to naming letters, not necessarily because we are unaccustomed to reading or recognizing them. Nevertheless Cattell concluded that in general we read by word-wholes and even sometimes by whole phrases at a time. Erdmann and Dodge supported this view because they found that a whole word could be read even when its individual letters were too far away or too small to be perceived separately and individually. They also stated that long words, particularly of an optically characteristic form, were more readily recognized than shorter ones. But this has been contradicted by Kutzner (89), who found that, other things being equal, short words could be recognized at a greater distance from the observer than long ones. Korte (88) has shown that short words can be more readily recognized in peripheral vision.

Goldscheider and Müller (58) concluded from their work on the tachistoscopic perception of diagrams, letters, syllables, words and phrases, that words are recognized principally by means of certain 'determining letters', of which the initial letter, letters projecting above or below the line, and vowels, were the most important. They considered that perception of these determining letters was followed by an auditory image or an actual vocalization of their sound, which in turn suggested the sound of the whole word; hence the importance of the vowels, in making the word pronounceable. The greater the familiarity of the reading material, the fewer the determining letters which must be perceived to lead to recognition of the words and phrases. Zeitler (157) came to rather a similar conclusion. He found that words which were exposed tachistoscopically for a very short time were recognized by means of certain 'dominating parts', which are equivalent to the determining letters of Goldscheider and Müller. The letters projecting above and below the line and certain other characteristically shaped letters are recognized preferably, i.e. they are the dominating parts. But in normal reading certain characteristic syllables or even words may become 'dominating complexes'; they alone are perceived, are arranged one after the other, and the remainder of the sentence or phrase filled in and completed by an inner mental contribution. With difficult and unfamiliar words, however, the reader is driven to perceive by means of dominating letters. The word sound is not necessarily an integral part of the process, but may succeed it. Huey (72) agrees on the whole with this view. He considers that visual perception of the dominating parts of words or phrases tends to arouse in consciousness the words most habitually associated with this partial perception. These associations are both visual and auditory-motor; but actual auditory-motor imagery of the visually perceived dominating parts, though frequent with some individuals, is not necessary to all. From the associated words and phrases which best fit the partial perceptions, the actual words and phrases are apperceived. These auditory-motor associations and images should be distinguished from the motorization of the complete phrase, known as 'inner speech', which lags at some distance behind the apperceptive processes.

Messmer (100), also working on tachistoscopic reading, concluded that readers could be divided into two types; the first, the objective, characterized by a rigid fixation, narrow perceptual span and objective accuracy of perception, and the second, the subjective, characterized by a fluctuating fixation, broad perceptual span and a tendency to interpret his perceptions subjectively. The first type of reader tended to perceive by means of dominating parts, as shown by Zeitler; and the second by means of the total word form. Messmer considered that the total word form is determined principally by its length and by its vertical profile, but also by its relative number of vertical letters (e.g. i, n, m, t, l, etc.) and curved letters (e.g. o, e, c, s, etc.). If the word is composed principally of one of these types of letters, it will be less readily perceived than if it contains a mixture of both types, or several vertical-curved letters (b, d, p, q).

Dodge (41) criticized Zeitler's and Messmer's use of a very short period of tachistoscopic exposure. He showed that under such conditions the assimilation of the word must largely take place from the after-image, since there is insufficient time for the 'clearing up' of the actual visual sensation; and demonstrated that any large or prominent features, such as dominating letters, would tend to be exaggerated in consciousness and remembered to the exclusion of all others. But in normal reading, or tachistoscopic exposures of 100σ and over, where, as a rule, the visual impression is fully cleared up, such letters would receive no undue prominence, and would stand out no more clearly than the rest of the word. Perception would thus take place by means of the total word form.

Kutzner (89) considered that the important factor in perception was the 'form quality' (Gestalt-qualität) of the word, which was determined partly by its length, but more particularly by the number of ascending and descending letters (letters extending above and below the line), and their position relative to one another and to the rest of the word. Thus when the word 'Vordergrund' was exposed, one subject guessed that it was 'Vagabund', because the word form seemed to contain the following ascenders and descenders, '*V*—ˌ—ˈ—ˈ'; at the next exposure he reported 'Vordergrund', on the basis of the word form '*V*—ˈ—ˌ—ˈ'. But these results were obtained from a series of tachistoscopic exposures of material which gradually approached the observer from a distance, a method of experiment the results of which cannot be applied without qualification to normal reading. The same may be said of the work of Korte (88), who studied the perception of words and letters seen in peripheral vision. He concluded that the 'total optical form' was not very important in ordinary reading. A vague total optical impression is followed by identification of the constituent parts, which arouse an auditory-motor image; and from these constituent parts a new image of the total form arises, influenced also by the meaningfulness of the word. He agreed with Messmer that words were more easily recognized if they contained a regular exchange of long and short letters, and were difficult to read if composed entirely of short letters. Ascending letters yielded better cues than descending ones. This view seems fairly similar to that of Goldscheider and Müller.

The work of Stein (134*a*), and of Heimann and Thorner (64*a*) (141*a*), has also stressed the importance of 'form quality'. The former considered that reading became more difficult the less favourable were the conditions for the effectiveness of 'form quality'. The latter found that 'total structure' was the leading factor in the reading of words; the factors which were of the greatest importance in contributing to this structure were the distribution of dominant letters and the pronounceability of the syllables in the words. (The difference in ease of reading between pronounceable and non-pronounceable syllable combinations was greater than the difference between the former and real words.)

In some work of the writer's (147) on the errors made in reading, it appeared that there was a tendency to omit letters in normal reading, especially from long words; thus with these words, the total word length was not of any great importance. The ascending letters were comparatively rarely omitted, the descending letters rather frequently, thus showing that there was a tendency to preserve the upper contour line of the word, but not the lower. This agrees with the results of Huey (71), who found that a passage in which the lower halves of the words had been deleted was read much more easily than a passage in which the upper halves of the words had been deleted. On the other hand, the writer found that these ascending letters, owing to their similarity of shape, were readily confused with and substituted for one another.

A word should perhaps be said in passing as to Huey's conclusion (71) that the first part of the word is more important for word perception than the second part. He found that material in which the second halves of the words had been deleted was read more rapidly and easily than material in which the first halves of the words had been deleted. It seems probable that this depends entirely upon the words used. In many of the shorter words, particularly those of Anglo-Saxon derivation, the 'root', the most important part of the word, is in the first syllable; it is only natural that if this part is removed, the reader is unable to apprehend the word. On the other hand, words of Latin derivation, especially the long words, frequently have the 'root' in the middle of the word, preceded by a prefix and terminated by a suffix; consider such words as 'inflammation', 'expectation', 'subservient'. Such words could only be apprehended from their middle syllables; while in the words from which they are derived, 'inflame', 'expect', and 'subserve', it is the last syllable which is of importance. Indeed, it was found by the writer (147) that educated subjects, reading material containing a large number of words of Latin origin with prefixes similar to those of the above words, though they made the largest number of errors in the middle of the word, yet made more at the beginning of the word than at the end. But the excess of errors in the middle of the word over those at the beginning and at the end was much greater for the less educated subjects, for the reason that they were unaccustomed to deal with these long words, and were very liable to confuse or omit their central and important part. Huey's results may perhaps be explained by the fact that most of the words in his reading material were short ones with the 'root', or important part, in its first half. The writer's data also show that the first letter is not necessarily a dominating or determining one, as postulated by Goldscheider and Müller, although it may be so in some words such as those just mentioned. The evidence for its dominance has been based partly upon the tachistoscopic reading of single words, where it is rather to be expected, and partly on the reading of German substantives which begin with a capital letter, naturally prone to dominate on account of its size and dissimilarity of type. Thus we may conclude that the dominating parts of words do not bear any invariable relationship to their total form, but vary with the particular word in question.

The conclusion seems to be that some general form or contour is perceived, with certain dominating letters or parts arising out of it, as the 'figure' rises out of the 'ground'. The ascending letters seem to play an important part, and an alternation of vertical and curved letters may also help in structuralizing the form. It is improbable that any individual letters or parts of words are recognized as such. But they are the details, standing out from the rest of the field, which differentiate its flat clearness, and finally produce perception of the 'specific object'. The words are not necessarily recognized and named individually, but the structuralized visual percept of the whole phrase or sentence arouses the more or less subliminal auditory or kinaesthetic imagery of the sounds appropriate to that percept, or in some cases an incipient vocalization of these sounds. This total percept is then interpreted directly in accordance with the general meaning of the reading content. In fine, we do not pass through complete letter recognition, word recognition, sentence recognition, and lastly apprehension of the general meaning of the sentence. But the vaguely contoured visual background, structuralized by the dominating parts to form a 'specific object', and the vague but differentiated auditory-kinaesthetic sensations or imagery based upon it, together are sufficient to convey to us the general significance of the whole perceptual situation—but this only, of course, after years of practice, and in a highly familiar perceptual situation such as the reading of easy prose.

The function of the general significance of the content is illustrated by the writer's work (147) on the type of error found in the oral reading of mature readers. Here the apprehension of the whole phrase or sentence was followed by the actual naming of the words contained in it. When wrong words were substituted, they were sometimes visually similar to the correct word, but did not fit in with the meaning of the sentence; sometimes they fitted in with the meaning, but bore no visual resemblance to the correct word; but more frequent than either of these errors were those which were both visually similar and also bore out the general meaning of the content. The partial visual perception was correctly interpreted, so far as the general significance of the content was concerned, but not with regard to the individual words which constituted it.

Evidence of the importance in the visual perception of the contextual setting on the one hand, and the general subjective basis of thought on the other, was mentioned in the earlier part of this chapter; it appears again and again in work on reading. Pillsbury(118), working on the detection of misprints in words exposed singly in a tachistoscope, found that the misprinted word was more readily perceived, and the error ignored, if some word of closely associated meaning was spoken just previously. Huey's (71) subjects, reading a connected piece of prose with each consecutive word exposed singly, frequently reported a 'forward tendency', or a 'tendency to fill out the sentence', which was strong throughout except at the beginning of paragraphs, when they were doubtful as to what was to come, and at the end of sentences, when there was no more to anticipate. Occasionally the exposed word conflicted with the one which had been expected on the basis of this filling out of the sentence. Indeed it is no uncommon experience in normal reading suddenly to come to a stop and find that one has been completing and developing one's partial perceptions along the wrong lines, supplying from one's own subjective thought processes an argument or a piece of description which sooner or later comes into conflict with subsequent incongruous partial perceptions. These thought processes which are carried on concurrently with reading are of the greatest number and complexity in mature educated readers. They may even determine to a great extent the words reported in tachistoscopic reading. Both Kutzner (89) and Pillsbury (118) describe cases of readers who had just been studying or reading books in foreign languages (French in the former case, French and German in the latter), and who reported that they saw words in those languages instead of those in the native tongue which were actually exposed. Similar cases of mental predispositions are given by many experimenters. Indeed, many readers, while glancing through a book or a newspaper, must suddenly have felt as if springing out at them, giving them a shock of familiarity, a word very closely connected to some mental preoccupation or strong interest. On further inspection they may have found themselves quite mistaken; the word in question was conjured up by the thoughts of the moment out of some partial perception of quite a different word.

Thus it appears that reading can in general be carried out accurately and efficiently on the basis of vague and rapid perception, provided that the individual is experienced in reading that type of material and is able to fill in the appropriate words and phrases. He must not, however, allow too free a play to the concurrent thought processes arising from his preformed interests and emotional dispositions; lest either assimilation of the reading content be altogether abandoned, or continual conflict occur between what is partially perceived and what is supplied in thought.

(3) The Perceptual Span

It is now of interest to consider the amount or rate of perception in reading. Clearly this is dependent upon the amount of material perceived at each fixation pause. This might be thought to depend in turn upon the field of clear vision. Now at the normal reading distance, the fovea could contain the optical image from a horizontal line of only 0·6 cm. in length, since it subtends a visual angle of about 70′; while the macula could contain the optical image of a line 6·0 cm. in length, since it subtends a visual angle of about 12° (Parsons(114)). It follows that at this distance only about four letter spaces of normal sized print can be seen foveally, although nearly forty can be seen in macular vision. At each fixation pause during reading a very small central area is clearly seen, and in the surrounding area vision must become more and more blurred towards the periphery. Ruediger(126) has mapped out the field of vision, by determining the distances from the central point, along various meridians, at which the letters 'n' and 'u' can be distinguished. The field has no clear boundaries, but fades off gradually. It varies in shape and extent with different individuals. Moreover, its size is not correlated with the speed of reading or the number of fixation pauses per line in reading. Thus the speed of perception in reading is independent of retinal sensory factors, and must be dependent upon central nervous processes.

Some light is thrown upon the nature of the perceptual span by the studies of tachistoscopic reading which have been made by Cattell, and by Erdmann and Dodge, as described on p. 111. These results seem to show that though the actual number of discrete objects perceived varied greatly from one individual to another, the ratio of the numbers of the different types of material were relatively constant. Messmer (100) considers, however, that the relative speed of perception of letters and words varies according to the type of reader, and his tendency to read words from their total form, or to analyse them into letters. And the results of some work by the writer (147) showed that the perceptual spans in the tachistoscopic reading of disconnected words and of words connected in sentence form did not necessarily correlate highly either for width or accuracy. The increase from letters to words, and so on, in the number of objects perceived seems to be due partly to the effects of grouping; but far more important are the effects of meaning and familiarity. The importance of these factors as an aid to perception has already been described. The effect of familiarity was clearly shown by the writer's work on the tachistoscopic perception of isolated words (147). There was a strong tendency to mis-read unfamiliar words, such as 'fend', 'atop', 'alum', etc. Certain such unfamiliar words were misread 5·2 times each, while the remaining more familiar words were misread only 2·8 times each. The effects of meaning are shown by the increase in the number of words read when they form a connected sentence. Another factor which makes for the rapid perception of words is the combination of the letters to make a single auditory and kinaesthetic unit, namely, the word sound. It was found by Zeitler(157) that four to seven consonants could be perceived at a single tachistoscopic exposure, but five to eight vowels and consonants interspersed. Again, pronounceable syllables combined into nonsense words of six to ten letters could be perceived at a glance.

It might be concluded that the perceptual span would be greater in normal than in tachistoscopic reading, since the scope of operation of the above factors would appear to be enlarged. Huey(72) cited cases of readers who occasionally made only two fixation pauses in the reading of a line of 9·8 cm. in length; their span covered nearly thirty letter spaces, and four or five words. This, it may be said, is a greater achievement than that of Cattell's reader (see p. 111), because in that case the words were arranged in two groups one beneath the other, and were therefore more favourably situated for rapid reading. But the achievement of Huey's reader was quite exceptional; the average span varied from five to fifteen letter spaces for his four adult readers. Most investigators have found that the normal frequency of fixation pauses per line indicates that one or two words are covered at each fixation; and this amount is frequently called the 'span of recognition'. Thus the whole available perceptual span is not normally utilized. This seems to result from the fact that the central processes of assimilation, and all the other mental processes concomitant with reading, require considerable time for their activity, and thus limit the rate of reading. If the amount of new material to be assimilated at each fixation is comparatively small, there is sufficient time for this assimilation to take place. It might of course be argued that the most suitable procedure would be to utilize the whole perceptual span and lengthen the fixation pauses sufficiently to allow time for the assimilatory processes. Some evidence was adduced in the last chapter to show that a tendency to adopt this procedure appeared in those individuals who were able to make long steady fixations and long accurate eye movements. C. T. Gray (59) found that the span of recognition could be increased even in fairly old children by suitable drill. This he attributed to the fact that a larger proportion of the maximum perceptual span was utilized, and there was less overlapping. Presumably also the rate of central assimilation was increased. He considered, however, that some overlapping was necessary in order that the thoughts and ideas aroused by one perception might be connected with those aroused by the next. Since the duration of the thoughts is not limited to the duration of the perception, this contention does not seem to carry much weight.

Hamilton (64), from her work on the reading of continuous prose by means of successive tachistoscopic exposures, concluded that clear perception and complete recognition do not necessarily occur within the overlapping areas, but that at each fixation there is an area of distinct perception, together with a set of marginal impressions situated principally on the right of the area of distinct perception. These impressions vary greatly in clearness. They serve as preliminary partial perceptions of words, and orientate the reader for the next clear perception. They are also the essential stimuli for the reflex eye movements. These observations seem to accord well with the deductions previously made as to the nature of the percept in reading—namely, that it is a vague, blurred, visual impression filled out with the help of a variety of thought processes which are concurrent with reading, and are based probably upon the reader's experience of that type of reading material, and that style and mode of expression.

The views of various writers seem to differ considerably as to the importance to the reader of a wide perceptual span, as measured in tachistoscopic reading. In any case, it is the perceptual span for connected words which must be considered, since, as was shown above, this need not be highly correlated with the perceptual span for disconnected letters and words; while the work of Whipple (153), Gates (52) and others has shown that it is unrelated to the ability to perceive other types of sensory material. Dallenbach (29) found that an improvement in the visual apprehension of young children was accompanied by a general rise in school ability. He agrees with Whipple (153) and Foster (48) that the only effect upon adults of practice in tachistoscopic perception is an increased adaptation to the experimental conditions; but holds that with young children a permanent improvement in visual apprehension may be produced by suitable methods of training with short exposures. He found that the average 'range of visual apprehension' before training was approximately two words, and after training approximately three. The effects of practice persisted over a considerable interval of no practice. These results he ascribed to the fact that the perceptual ability of children is less mature than that of adults. Hence, as would be expected, children whose ability was originally poor improved more, and over a longer period, than those whose ability was originally good. Even with mentally deficient children a very slow and gradual but permanent improvement may be produced, which also seems to increase their mental age (30). Gray (59) found that what he terms the 'span of attention' of young children in the tachistoscopic reading of continuous prose could be increased by practice with short exposures and rapid reading, and that the rate of normal reading was also increased. He considers that differences in the 'span of attention' are not inherent, but are due to the acquired ability to deal with meaning; and also as to whether small or large units of perception have been employed in learning to read. The inter-relations of perceptual span, intelligence and reading ability seem to be confirmed by the results of Pyle (122*a*). He measured the reading span by finding the longest sentence of easy words which could be read during a tachistoscopic exposure of 3 secs. This span increased continuously from the ages of (roughly) six to fourteen. Moreover, it ranged from 4·5 to 13·8 words for the most intelligent of these children; from 3·3 to 13·2 words for those of medium intelligence; and from 2·9 to 11·5 words for those of poor intelligence. College students could read 15·6 words. Again, Pyle found that if children aged about twelve years were divided into five groups according to their reading ability, the reading span ranged from 7·3 words for the poorest group to 12·4 words for the best.

The use of these various terms, perceptual span, reading span, span of attention (which seems to correspond to perceptual span), range of visual apprehension, makes it difficult to be certain just what these workers have studied. Moreover the number of factors underlying visual perception and apprehension have been shown to be so diverse as to make it impossible to attribute the width of the perceptual span to any one of them. Familiarity with the symbolic letter and word forms, and with the phrases and verbal constructions into which these words can be grouped; a wide basis of thought and experience to which the ideas suggested by these symbols can be connected; the general factors, attitudes, moods and interests, based upon a temperamental foundation, and general intelligence; all these factors may contribute to a wide perceptual span in tachistoscopic reading, just as they contribute to a wide span of recognition in normal reading. Thus any form of practice or method of training which effects some growth or improvement in one or more of these factors may widen the perceptual span in tachistoscopic reading, and increase the span of recognition in normal reading. In so far as the so-called 'flash-card drill',[1] consisting of practice in reading large

1 'Flash-card drill' is usually accompanied by a number of other intensive methods for improving reading ability; hence it is difficult to attribute improvement to any one of them. Indeed, it may often be due simply to the increased care and attention devoted to reading.

groups of words exposed momentarily, increases the ability to group words into large perceptual units, so far does it seem possible and even probable that the perceptual span will be widened, and the rate of reading increased. It does not appear, however, that such practice need in any way affect the mental processes connected with the apprehension of meaning and its assimilation to the body of previous thought and experience; not to mention the variety of subsidiary processes of imagery and associated thought, which are frequently necessary to give the reading experience its full value. These can only come by growth and training in thought and understanding.

(4) Variations in the Nature of the Perceptual Processes in Reading

It must be remembered that the rapid and partially perceptual type of reading does not invariably take place; it has not been acquired by the immature or uneducated reader. Moreover, with difficult or unfamiliar material, it may sometimes be abrogated and replaced by a much slower type of reading which includes a clearer and more complete perception of the individual words. It seems unlikely that letter for letter reading is ever resorted to except in the case of unknown or foreign words. But as Zeitler (157) pointed out, it is not improbable that the dominating parts of single words are perceived individually and that the associative processes described by Huey (see p. 113) arise only from single words or very small groups of words. According to Messmer (100) the objective type of reader is naturally inclined to read rather from the dominating parts of the words than from the total word or phrase form, and is also characterized by a narrow perceptual span and objective accuracy of perception. But it was shown by the writer (149) that a narrow span of recognition in reading, as indicated by the frequent short fixation pauses, was often accompanied by a large number of regressive movements and refixations, which seemed to denote inaccuracy of perception. Moreover, this narrow span was correlated with lack of accuracy and steadiness of voluntary movement and fixation. Thus it may be said that a habitually narrow span of recognition is a function of lack of ocular motor ability, and need have no connection with the tendency to apperceive the words by means of their dominating parts. This is supported by the fact that some of the individuals characterized by this narrow span reported a large amount of subjective elaboration and associated thought which, according to Messmer, should be characteristic of the subjective type of reader, with wide perceptual span.

The view that words or parts of words are perceived individually with difficult and unfamiliar material, is supported (though not proved, since visual fixation is not necessarily coincident with perception and assimilation) by the observations of Judd and Buswell (82), described in the last chapter, on the increases in the number or duration of fixation pauses which were shown to occur if the reading material was rather difficult or unfamiliar, or if the reader was asked to read carefully. We may suppose that a moderate amount of regression and refixation show that the attempt at filling in the partial perception has failed, and that a second attempt is necessary for apprehension. But Judd and Buswell also observed that in the reading of very difficult and incomprehensible material, or during the analysis of the text for purposes of paraphrase or translation, there occurred much wildly confused movement and regression, which did not seem to indicate an increase in the clearness and accuracy of the perceptual processes. Such a clear perception of all the letters, or at least the dominating parts of the words, would not be obtained by oscillating from one word to another, but rather by pausing long and steadily on each word in succession, a type of behaviour which rarely occurs in such circumstances. From this we may conclude that the perceptual processes do not become much clearer, but must rather be greatly confused and thrown out of order. Indeed it seems that with material sufficiently difficult and strange we are liable to 'make confusion worse confounded' by trying to utilize our normal habits of rapid partial perception; when these break down we are unable to read word by word, as the beginner does, but must drift aimlessly to and fro, picking up perceptions and still trying to fit them into a sensible whole. This occurs partly because we are unable to free ourselves from our accustomed habits; but partly also because both motor and perceptual processes are thrown out of gear by the direct influence of conative impulses which affect it adversely as they affect so many other habits of behaviour. For it was found by the writer (149) that wide variations in reading time per line and a large number of regressions appeared, not only when the reader was struggling to understand what he was reading, but also when some strongly affective reaction was aroused or when the material set up a train of thought organized within some strong personal interest. Such affective reactions and conflicts between interests and apprehension of the meaning of the content were believed to be accompanied by strong conative impulses, as were the efforts at comprehension also. Thus we may perhaps conclude that the impairment of perceptual and assimilatory processes by irregular and fluctuating eye movements results from the arousal of such conative impulses.

On such occasions it is the relationship between the motor and perceptual processes which is impaired. But on other occasions the motor and perceptual processes may retain their close integration, while the assimilatory processes are relegated to the fringe of consciousness. Thus it was found by the writer (149) that very regular and increasingly rapid reading occurred when the reader was scarcely attending at all to the content, and was unable at the end to recall it. This regularity was largely due to deeply rooted motor habits. But the movements were guided and regulated by some form of perception, for they were properly adapted to the phrase and sentence divisions of the text. Moreover, the perception of an unusual or incongruous word would sometimes break in upon the individual's train of thought and he would become aware of the incompatibility of the latter with the reading content. Thus it appears that in the mature reader the perceptual processes have become so habitual and stabilized that they may even proceed regularly in the absence of the associative and assimilatory processes which were originally necessary to their efficiency and rapidity.

Chapter VI

THE VISUAL PERCEPTION AND READING OF CHILDREN

(1) The Visual Perception of Children

We have dealt with the method by which the mature adult reader perceives the material which he is reading; and it is now necessary to give some account of the way in which that method develops and matures. It is a matter of great difficulty to discover what the young child perceives when he reads, since few indications can be obtained from introspective evidence. Moreover, although a vast amount of experimental work has been carried out on the various formal methods of teaching reading in schools, detailed work on individual cases is scanty, often inaccurate, and almost of necessity uncontrolled. Thus it is of interest to consider first the experimental work performed on children's perception in general; much of this seems to be careful and accurate.

In this experimental work the child is usually presented with a series of conventional forms, or more or less complex pictures of real objects or events. He may be required to match the simpler forms or objects with each other, or to differentiate them from each other, by means of form or of colour or of both. Or with the more complex designs or pictures he may be required to describe and name all that he can. A variant of the latter method is called the

'Aussage' experiment. A picture is shown to the child for a short time only, and then taken away; the child is then required to describe all that he can remember of the picture. These recollections may be repeated subsequently at varying intervals. Or sometimes a number of objects is exposed instead of a picture. It may be remembered that this variation, called the 'jewel game', was tried by Lurgan Sahib with *Kim*. But these latter methods, of course, involve memory as much as perception.

The first and most important result, substantiated by several experimenters, which emerges from this work is that accurate perception of form, particularly of conventional form, is poor in the young child. Thus Descoeudres (34) required children to match conventional forms and drawings of real objects; they could match according either to colour or to form, since the colours and forms were differently combined in the different designs. She found that children from three to six years of age matched predominantly according to colour, those from seven to thirteen about equally either way, and those over thirteen chiefly according to form; three-quarters of the adult observers matched according to form. Segers (130) obtained similar results. He found that colour predominated over form with conventional shapes up to the sixth year; but with representations of real objects colour predominated over form only till the fourth year. There seemed to be a sharp change in this attitude towards conventional form between six and seven years of age, when the child went from the kindergarten to the primary school. Again Segers found that visual perception of differences of form was poor in children of four to five years, and was not really reliable till ten years. The existence of dissimilarity between two forms was perceived at a younger age, however, than that of similarity. Line (92a) also found that recognition of conventional or abstract form was poor in early years, although a square could usually be differentiated from a circle, and perhaps from a rectangle. Some appreciation of relationships and ratios of size and shape began to appear from four to five years, however, but very little appreciation of relationship of colour. In general, the development of the perceptual cognition of abstract relationships was progressively selective. Stimuli at first had to be intense, massive or moving to be cognized, but could become increasingly fine, until minute details were perceived.

This inability to differentiate and compare by means of form probably results from the childish tendency to see objects as a whole, called by Segers 'syncretism'. Thus he showed children a number of drawings of animals in which the head of one animal was combined with the body of another. He found that no child of less than seven years, and few of less than nine years, appreciated this complexity. Nor was there any hesitation in naming the animal. Up to five years the children named principally according to the body and general appearance; and after that principally according to the head. Koffka (85) would interpret this by saying that the sensory configurations of children are so firm that they can only be altered as a whole. If a single part of the stimulus object is changed, this alteration may be unnoticed; or the whole configuration will appear to be altered, in direct opposition to the objective conditions of stimulation. But an apparently more fruitful and profitable explanation is that there is always a tendency to approximate any form represented to some real, familiar and interesting object. In the above experiments, it must be deduced that these children saw complete representations of animals because they were interested in animals, and were not interested in any imperfection in the representation. This appeared also in the work of Parsons (113) on children's interpretations of ink-blots; she found that over half the ink-blots which she presented were interpreted as being representations of human beings or animals.

On the other hand, if the design or picture presented is too complex to be interpreted as a single and simple whole, the various parts of it will be enumerated singly and separately, without the deduction of any relationship between them. This was illustrated by some of the children, aged about seven, who carried out Parsons's experiments. The parts of the blots were given separate and unconnected interpretations. Binet (11) found that with complex pictures there was enumeration of objects at three years, and description of the picture at seven years, but no interpretation of the picture as a whole till fifteen years; Segers (130) found that there was simple enumeration from three to five years, and complex enumeration, with objects occasionally related to each other, from five to six years; after that description, particularly of actions, was employed, and unimportant details were relegated to the background. Interpretation, when the subject of the picture was understood and described, did not begin until eleven years. Stern (135) called the first stage the 'thing' stage, and found that it continued up to seven or eight years of age; from eight to nine was the 'action' stage, when various activities were described, and after that the 'relational' stage, when the parts of the picture were logically related to each other, and a general interpretation supplied. There seems to be some difference of opinion as to the age at which the 'thing' or 'enumeration' stage is abandoned. But as Segers (130) points out, this may very well vary with the type of picture presented. The young child cannot understand the general meaning and significance of a complex picture, and is not interested in such a general meaning. Hence it picks out separately any objects of interest which attract its attention. It follows that the more closely the topic of the picture and the activities represented in it are related to the child's experience and interests, the younger the age at which some kind of general description and interpretation will occur. It seems probable that action of some kind will be more readily appreciated than merely static relationship.

There seems to be some evidence that the differences in the type of perception do not merely represent stages of maturity, but that some types may predominate more in some children than in others. Thus Binet (10) deduced from children's descriptions of pictures the existence of four perceptual types: (1) the descriptive, who observes and enumerates objects in a picture without searching for any significance;

(2) the observational, who notices attitudes and expressions, and the general topic of the picture; (3) the emotional, rather similar to the observational, who reports the emotional interpretation, but gives less descriptive detail; (4) the learned, who is quite impersonal, and describes only what he is sure of. While the first two types clearly correspond to the 'thing' or 'enumeration', and the 'interpretation' stages, the third and fourth seem to resemble more the adult 'subjective' and 'objective' types of Messmer (100). This adult classification was also adopted by Piaget and Rossello (117), who obtained written descriptions of three pictures from forty-three children of eight to twelve years, and followed them up by individual examinations. They found (1) the objective type—matter-of-fact and accurate, and (2) the subjective type—imaginative and inaccurate; but also (3) the intelligent type, who combined the qualities of (1) and (2) with the deduction of explanations and hypotheses, and (4) the superficial type, who showed neither accurate perception nor imagination. While (1) and (2) resemble the main adult perceptual and possibly temperamental types, (3) and (4) seem to denote large differences of general intelligence combined with these perceptual and temperamental differences. However this may be, these authors certainly consider that these types represent general attitudes based on the temperament or personality. Frank Smith's work (132) on the tachistoscopic perception, in a long series of exposures, of complex pictures by children of six years, seems to show that these differences may be the result of the combination of innate intellectual factors and of

the environmental factors which differentiate children drawn from different social classes. Thus he found that children from an elementary school in a poor district were very inaccurate in their observation and showed a great tendency to invent imaginary details, both in accordance with association to previous experiences, and also as the result of suggestion by the experimenter. Smith attributed the latter trait in particular to the fact that they were in the habit of giving any reply likely to be acceptable to the 'powers that be', and were relatively unconcerned with its truthfulness. It may also have been a function of poor intelligence. Children of the same age from an elementary school in a better district of the same town were much more accurate, invented less imaginative detail and showed a generally more critical attitude to the pictures presented.

Segers(130) considered that the relative objectivity and subjectivity of the manner of perception was merely a function of age. He stated that the objective type was much more common in children from six to eleven years; while the subjective type became increasingly common from nine years upwards. Messmer(100), however, considered that children on the whole conformed to the subjective type of behaviour; that is, in the tachistoscopic perception of words and letters they showed a fluctuating fixation, broad perceptual span, and an inward and subjective interpretation of what they perceived. Smith(132) also found that children of twelve years, especially those of superior education and intelligence, were more objectively accurate than children of six years, and on the whole were less ready to interpret the story contained in a complex picture, although they often inferred the existence and position of other objects in the picture from the data afforded by previous perceptions of the picture. Jaensch (73) found that his eidetic type was more common until the age of eleven or twelve years than afterwards. The eidetic was characterized by an intimate connection between perception and eidetic imagery, the two mental functions following virtually the same laws. Thus perception was essentially plastic, and closely integrated with purely subjective mental functions on the one hand, and with bodily functions, such as thyroid secretion, on the other. This type of child was naturally interested in real objects, representations of actual familiar events, etc., and would tend to interpret his percepts freely in accordance with his own subjective interests. The opposite perceptual type, characterized by objective accuracy and lack of subjective interpretation, was also found among children, but was much more common after puberty. Thus the differences in the theories of the various experimenters might be explained by the differing incidence of eidetics and non-eidetics among the children they observed.

The general deductions seem to be as follows. Presented with any perceptual object, the young child's response will be given in accordance with the aspects of the situation which appeal to his interests and are within his comprehension; these exclude minute details of form and structure. If the stimulus is simple, he will interpret it as some familiar object. If it is complex, he will report all the objects, activities and situations in it which are familiar to him, and may invent others which are familiarly coupled in his mind with those actually seen. If he is very young, the general significance will probably be beyond his grasp; but at a slightly greater age he may show considerable powers of interpretation and subjective elaboration. Also, if the stimulus object contains comparatively little of interest to him, as does, for instance, a single word or a collection of letters, his eye will rove round looking for something interesting, and ultimately he may invent some such thing if he cannot find it. Thus although the type of response given may vary according to age, temperamental type, intelligence and environmental conditions, it is probably not a direct function of any of these diverse factors, but is connected with the varying interests and familiarities which are themselves functions of these factors. Thus the types of perception appearing in the young child are predominantly of a subjective nature, in that the part played by the objective stimulus is secondary to that of familiarity and interest. It seems probable that the extremely accurate objective type definitely shows a highly developed form of response, although in Jaensch's disintegrate type the development may have occurred phylogenetically and not ontogenetically. The subjective or the integrate type may, however, survive in many cases to the adult stage.

(2) Children's Reading

It seems to follow that children under six years of age will have great difficulty in learning to read, since they will be unable or unwilling to perceive the small dissimilarities of form which differentiate letters and words from each other. However, Segers (130) observed that there was much improvement in the perception of form at the age at which the children went from the kindergarten to the primary school; that is to say, presumably, when they began to learn to read. Thus it may be deduced that if the child is sufficiently anxious to read, he will learn to differentiate the visual percepts which are aroused by the printed words. It follows that reading will only be learnt easily if (*a*) the meaning and significance of the reading material are comprehensible, familiar and interesting to the child, and (*b*) if the words used are commonly found in his vocabulary. The importance of the second factor has been stressed by Wiley (156). He pointed out that Thorndike's list of words to be taught to children had been drawn up on the basis of words most commonly used by the adult, which words may have quite a different degree of utility in the child's vocabulary. He found that the child's richness of association with certain words, as measured by the quickness of response in a free association test, correlated with the ease of learning to read them to the extent of $0{\cdot}55 \pm 0{\cdot}06$. It may be deduced that words of apparently simple perceptual form, for instance, certain monosyllabic words such as 'axe', 'lad', 'mit', may be quite unsuitable for teaching to beginners, because they are unfamiliar to most children.

In accordance with the first of the above principles, some writers have advocated that the teaching of reading should begin with simple sentences of about four words each. A sentence will probably mean more to a child than a single word, particularly if it deals with some familiar activity. Brown (18) considers that such a sentence can be learnt as an unanalysed visual whole, just as a spoken sentence is originally learnt by a child as a group of sounds with a unitary meaning and significance. When several sentences have been learnt, single additional words should be taught, and hence analysis of the sentences into words. This type of procedure was adopted by Mosher (104*a*). The children invented themes of about four sentences each, dealing with their own home-life and experiences, in their own words. These the teacher wrote on the blackboard and read to the children, who then repeated them until they knew the words. This process was carried on from day to day, and the individual words re-tried and tested. Finally books were introduced; the children read them with ease and interest, and were able to make out new words to a great extent by inference from their context. It was found that this method was most successful with the more intelligent children, who in the course of a year acquired a large reading vocabulary; but it was much less successful with the less intelligent, who presumably required in addition some more analytic method. Further experimental support for the sentence method of teaching is given by the work of Boggs(14), who taught children to recognize letters, syllables, words and short sentences written in Greek script. She found that all the children recognized the sentence units more often than any other unit, and after that the word (except for one child who had learnt the English alphabet; he associated the Greek letters with the English ones, and hence recognized them more easily than words).

But even if at the beginning children are taught to read in sentence units, it is clear that all of them, except possibly a few of the most intelligent, must pass through some stage of analysis before they can finally attain to the adult method of perception of phrases and sentences. Otherwise they would be at a loss in dealing with any unfamiliar phrase or sentence. Similarly they must at some time learn the analysis of the word into letters, in order to be able to deal with new and unfamiliar words. Bowden(15) was of opinion that the child must be taught to make this analysis, it being unlikely that he would do it for himself. This indeed seems probable, for the child would have no natural interest in the structure of the word apart from its meaning. To deal first with the reading of words as units. The work of Boggs(14) showed that the word unit was, other things being equal, more easily learnt than the syllable or letter. Again N. B. Smith(133) found that even children who had learnt their letters recognized them with more difficulty than they did larger units. In matching letters there was much liability to confusion between letters of similar form, such as 'b' and 'd', 'p' and 'q'. But nonsense syllables were much less readily confused, even when they contained these difficult letters. And Sholty(131) found that, among three children, the best reader always saw the total word form, and never analysed it into parts. But she made more mistakes than the other two readers who, less good at reading because they had difficulty in recognizing words at sight, were yet more reliable. Gates (54) also found that children who had been taught to read by a word-whole method were markedly superior in silent reading comprehension to those who had been phonetically trained.

But clearly there must be some feature in the total word form which differentiates it from other forms. The work of Bowden(15) seemed to show that children in learning to read individual words paid little attention to the total word form or general contour. Thus words were read without difficulty when presented upside down, or when the letters of the word were transposed, as when 'nettims' was presented instead of 'mittens'; but alteration of letters with preservation of the word contour, as in 'lihac' for 'lilac', was immediately detected. Again, although the most common confusions were between words of the same length, confusions due to identity of certain letters were also very frequent. Structuralization of the total word form by means of ascending and descending letters, or an alternation of straight and curved letters (see p. 114), did not help the children to read the words. Contrary to the opinion of Messmer(100), 'linear' words, without ascending and descending letters, were read more easily than others by four out of six children; while all the children read words consisting principally of straight line letters more easily than any others. But there was no definite analysis of the word to pick out familiar letters or letter combinations. It must be concluded that reading was a rather imperfect form of word perception and recognition depending upon the recognition of the presence or absence of certain familiar letters.

The work of Gates and Boeker (55) also seemed to show that recognition of individual letters was more frequent than recognition of total word form. Thus it was not noticeably true that words of very similar form were confused by beginners. 'Hen' was confused with 'man' and 'box' as often as with 'pen', and 'pen' was confused with 'pig' and 'rat' as often as with 'hen'. 'Boat' was called 'drum', 'bird' and 'kite' more frequently than 'foot' which it clearly resembles. There was a tendency to confuse words with similar endings, such as 'horse' and 'purse', but similarities in the beginning or middle had no noticeable effect. Wiley (156) stated that a large proportion of errors in children's reading was due to similarity of configuration; he found a correlation of $0{\cdot}86 \pm 0{\cdot}02$ between total errors and errors due to this cause. But these confusions were due to identity of letters rather than to general similarity of the total word form. Thus confusions due to similarity of beginnings and endings were about equally frequent. Errors due to the ending of one word being similar to the beginning of another, as in 'girl' and 'dog', were rather less frequent; while confusions between words with similar middles were comparatively infrequent. Again Meek (99) found that if words which were being learnt were presented together with five others which had one or two letters the same (e.g. 'ball' with 'burr', 'feel', 'sale', 'bake', 'kill'), the most confusion was caused by words with the two final letters the same and less confusion by those with the first two letters the same. On the other hand, those with the same first letter only were more readily confused than those with the same final letter only. She concluded quite definitely that certain letters or groups of letters were used as 'cues' for word recognition; 'i', 'g', 'll', 'o' and 'k' seemed to be favourite ones. But no hard-and-fast rule could be laid down as to which letters would be selected; they differed according to the words in which they occurred. Thus 'doll' and 'bill' were often confused on account of the 'll', but 'kill' and 'ball' comparatively seldom. Again, younger children tended to confuse 'flag' with 'drag' on account of the 'ag'; but the elder children confused it with 'fled', on account of the 'fl'. Gates and Boeker (55) found that words were often recognized by special details, such as the dot over the 'i' in 'pig', and 'funny cross' at the end of 'box'.

One must conclude that children taught by methods such as those of Meek, and Gates and Boeker—that is, by learning to pick out certain words from among a number of others of similar form—recognized words from certain individual letters or letter combinations, and not by means of the contour or structure of the total word form. Had they been taught to differentiate between words of similar contour and structure (containing the same number of straight and curved, ascending and descending letters) with different actual letters, it might have been possible to show whether word recognition in such a case depended rather upon contour than upon the presence of certain prominent letter 'cues'. Gates (54) in an extensive series of experiments compared the results of teaching reading by phonetic methods, and by non-phonetic methods in which the words were presented as wholes in context; that is, no particular stress was laid upon the differentiation of these words from others containing many of the same letters. He concluded that the latter method improved the perception of the characteristic total configurations of words, and produced rapid appraisal of these words. Phonetic training centred attention upon the smaller details of the words, particularly upon the two-letter phonetic elements, and resulted in a detailed scrutiny of the words presented. Thus it appears that the reader will have a bias towards perception of word-wholes or of certain letter 'cues' according to the method by which he was first taught to read. Judging from the experimental work of Sholty (131), Gates (54) and others, the word-whole method seems to be the more successful. But it must of course be succeeded by some process of analysis.

Again, it appeared that no definite conclusion could be drawn as to which letters, in what positions in the word, were employed as cues for perception by those who made use of these cues. As Meek (99) pointed out, they varied from word to word, and also, probably, from reader to reader. The only general conclusion was that the initial and final letters were recognized and remembered better than the middle letters of the words. Nor does it seem possible to deduce which type of word will be easy to learn, and which will tend to be confused with others; although it is reasonable to accept the findings of Gates and Boeker (55) that, other things being equal, short words are more easily learnt than long ones. But here there is very little doubt as to the importance of the part played by the different associations connected with the words. It is impossible to allow for or to avoid this qualification, since no one can tell exactly which words will be most familiar and interesting to any particular child. Thus it was perhaps unjustifiable for Gates and Boeker to state that the words 'pig', 'foot', 'cigar', 'basket' and 'fireplace' were not more familiar and interesting than the words 'hen', 'boat', 'knife', 'hammer' and 'telephone' respectively; and to conclude, from the fact that the former were read correctly more often than the latter, that ease and difficulty of reading are not dependent upon such associations. Wiley's (156) utilization of free association might perhaps provide a method of assessing this factor. To study perceptual form without taking these subjective associations into account does not appear likely to lead us any further in the discovery of what the child perceives when he reads, and what particular confusions of letter and word form make reading difficult for him.

It should not be deduced from the observation that phonetic training seemed to be inferior to non-phonetic training that the visual impression of the word or sentence should be divorced from its equivalent in speech. It was shown in the last chapter that association of the word sound was an integral part of the reading process. The visual percept must become integrated with the corresponding sound in speech in order that its meaning may develop from the meaning and associations of the spoken word. Thus the feeling for language relationship will mature concurrently in the spoken and written words and phrases. The difficulty of learning to read without the appreciation of the word sounds corresponding to the visual percepts is shown by the poor reading ability of deaf children, who cannot make use of the word sounds, though their ability to perceive words visually may nevertheless be high (Gates (53)). But once the visual percept and the word sound have become completely integrated, so that the former immediately suggests the general significance and also the language relationships of the latter, the way is open for the organization of the visual percepts of individual words into structural wholes, namely the habitual language and thought units of the phrase and the sentence.

(3) Reading Disabilities

A large number of cases are encountered of children who apparently are unable to learn to read by the ordinary methods. A considerable proportion of the disability is the result of mental deficiency, or such physical defects as poor vision or hearing, or defective speech. But when these cases are set aside there remain others whose disability is difficult to diagnose and remedy. Many such cases have been described, principally by American psychologists. But it is rare that any systematic and comprehensive scheme of psychological diagnosis has been adopted, and in consequence it is not known if these defects are due to hereditary or innate causes, to general environmental conditions, to the method of teaching reading, or to different causes in different cases. The disability is frequently removed by remedial treatment; but many different forms of treatment are generally employed at the same time, so that cure can be ascribed to no one of them in particular, and no light is thrown upon the cause.

A comprehensive diagnosis of fifteen cases of reading disability in children has, however, been published by Hincks(65). She found that all the cases had poor visual and auditory memory for words as wholes. Thirteen cases out of fifteen had some additional visual or auditory disability of a psychological rather than a physiological nature. Among the visual disabilities were lack of ability in the perception of forms of a certain character in the appropriate way. Thus some children had a good memory for outlines of large simple geometrical forms, and could perceive words vaguely and inaccurately as wholes; but they had no memory for small details, and thus were unable to correct their vague perceptions of total word form by attention to smaller details of dominating letter cues. Others had the opposite type of disability. They had a good memory for small details, but a bad one for large simple forms; consequently they could perceive and differentiate small parts of words, but could not synthesize them into total word forms. These disabilities may perhaps have been the exaggerated consequences of teaching in the first case exclusively by means of word-wholes, without supplementation by any more analytic method; in the second case, by means of letter and phonetic units, with no attention to synthesis. Hincks also found that the majority of these children had very limited peripheral vision, which she considered might curtail the visual, and hence the perceptual span, and force them to read in very small units. It seems improbable, however, that the visual span would be so small as to limit the reading span; the latter could probably have been greatly expanded had other disabilities been removed. Indeed, Anderson and Merton (3) found cases of reading disability characterized by very narrow spans of recognition in reading, which they definitely attributed to undue stressing by the teacher of 'word-drill'; thus the child had learnt words only as isolated units, and was unaccustomed to reading them combined into meaningful phrases. Again, eye movements and fixations, both among Hincks's and Anderson and Merton's cases, were usually irregular and uncertain, but these were improved by remedial training, and were also, in all probability, a result rather than a cause of reading disability. Freeman (49) and W. S. Gray (60) also have described cases of reading disability, prominent features of which were irregular and wandering eye movements. Freeman attributed them to extensive phonetic drill, which had caused the child to try and read words from the sounds of individual letters or letter groups, without any attempt to synthesize them into meaningful wholes. Thus the irregular eye-movements were the result of wandering about trying to pick out the familiar sound groups. The case described by W. S. Gray had no apparent defect in his visual memory; but his very irregular and wandering eye movements may also have been due to a tendency to read in small units, without attention to the meaning of words and content.

Such auditory defects as poor pitch discrimination were frequent among Hincks's (65) cases. The children could not hear the difference between phonetic sounds, and found vowels particularly difficult. Poor discrimination of pitch seems sometimes to be an innate defect, as in the tone deaf; but it can very often be improved with practice.

Others workers (Bronner (17), Lord, Carmichael and Dearborn (93)) have found reading disability to be centred in the associative processes by which the perceptual, auditory and kinaesthetic factors are synthesized and related to the meaning of the words and phrases. This disability in synthesis might be due to (*a*) poor mentality (short of actual mental deficiency), (*b*) emotional and temperamental defects, (*c*) some specialized defect which only appears in reading, or (*d*) simply the fact that the child is uninterested—he either does not want to read at all, or does not want to read the material which is given him. But that there is some real specialized defect in the synthetic processes seems to appear from the success gained by the remedial work of Fernald and Keller (47). These workers teach words to children by making them trace and then write the words many times over, saying them meanwhile without looking at the copy. The kinaesthetic sensations set up by the hand movements in writing seem to provide a link or mode of synthesis between the visual memory image on the one hand, and on the other the auditory-verbal sensations and the meanings associated with them in speech. It is difficult to diagnose the nature of the reading disability from this remedial method; unless it be that verbal-kinaesthetic sensations and imagery are weak or distorted, and must be reinforced by kinaesthetic sensations from the hand and arm. Also, it is clear that this method has the advantage of preserving the word as a unit of perception, speech and meaning, while enforcing attention to the details of its visual structure. In this it has a great advantage over the phonetic method, advocated by Lord, Carmichael and Dearborn (93) for remedial work in reading, since this divides the words up into small sound units and minimizes the importance of the word meaning. It might be useful for cases, such as those encountered by Hincks (65), who habitually treated the words as wholes, and were unable to analyse them. But even then the method of Fernald and Keller seems superior because it stresses the details of the visual form while preserving the unitary nature of the word sound.

Apart from those cases in which disability can definitely be ascribed to the undue influence of some one method of teaching reading with the exclusion of all others, it is usual to find several different types of defect in each case of reading disability. Thus many of Hincks's cases had poor ability in form perception, limited peripheral vision and inefficient eye movements, which might or might not be allied with auditory disabilities, and sometimes also with kinaesthetic disabilities. In addition, temperamental defects and emotional instability are very common. These difficulties may of course be the result of reading disability, or rather, of the feelings of inferiority aroused by the latter and their various compensatory reactions. But there seems to be some evidence that the reading disability is the result of definite neuropathic tendencies. Hincks found that the majority of her cases had emotional difficulties, and in many of them neuropathic tendencies appeared to be hereditary. Hollingworth (67) considered that neurotic children were often deficient in reading ability; they were unable to co-operate, to adhere to definite directions and to make sustained efforts at learning to read. She was of opinion that there is no such thing as specialized reading disability, but that any child of normal intelligence could be taught to read, if given special training in overcoming his particular difficulties, and also special attention to his emotional instability. It is quite possible that some temperamental deficiency might be found in every case of reading disability. On the other hand, temperamental deficiencies occur in a very large number of cases where there is no such disability. Thus even if they are a predisposing condition, some other direct cause must usually be sought. Bad teaching and lack of any attention to individual difficulties may account for many cases. The writer is inclined to think that uninteresting material and an unfamiliar vocabulary may have convinced the child at the beginning that it was not worth while trying to learn; and that the symptoms which have been described above developed simply as defence mechanisms or hysterical manifestations which thereafter prevented reading. There is no definite evidence on this point, apart from the importance of the part played in expert reading by interest, familiarity and comprehensibility of the reading material.

But two other theories must be described. The first is that of Hinshelwood (66), who attributed reading disability to 'congenital word blindness', that is, to some hereditary or innate pathological condition of the visual memory centre in the cortex. He based this theory upon the fact that he had observed 'word blindness' or 'alexia' to be produced by lesions of the brain which affected a particular area of the cerebral cortex in the angular gyrus. But such cases were re-educable only to a very slight extent, whereas children with reading disability alone can usually be taught to some extent by the use of suitable methods. This Hinshelwood attributed to education of the opposite hemisphere. When the child could not be taught to read, it was usually suffering from general mental defect, which Hinshelwood considered would affect both hemispheres. But no authentic cases have been demonstrated of definite organic defect which affected the reading processes alone. And even if such cases should occur, they are far too rare to account for all the numerous cases of authentic reading disability.

Orton (111) (112) has attributed reading disability to a purely functional defect of the brain. His theory is based upon the following features commonly observed by him in these cases:

(1) Confusion of the letters 'p' and 'q', 'b' and 'd', and of reversible words such as 'was' and 'saw', 'not' and 'ton'; also a tendency to reversal of pairs of letters or of syllables in words.

(2) Capacity to read and write mirror writing. This facility in dealing with mirror reading and writing was also found by Monroe(104) among retarded readers.

Orton explains these facts as follows. The two halves of the brain are equally stimulated by visual impulses in reading, but these stimulations are normally stored as visual memory traces or 'engrams' in one hemisphere only. This is shown by the fact that the ability to read is destroyed if the dominant hemisphere is injured, but not if the other hemisphere is injured. Thus the nervous units of the dominant hemisphere become the controlling ones, but the activities of those of the other hemisphere must be suppressed by elision of the 'engrams' set up by the visual impulses. If this elision is not complete, the persistence of these 'engrams' will result in a confusion between normal perceptions and their mirror images; mirror writing and reading will be facilitated, and also a tendency to confusion of reversible letters and words. This defect Orton called 'strephosymbolia'. It is particularly likely to occur when pure sight-reading methods of teaching are employed, without phonetic drill; and Orton states that a larger percentage of children with this disability is found in schools where sight reading methods only are employed than in schools where other methods are also utilized. It can be remedied by a method such as that of Fernald and Keller(47), in which the word sound and its kinaesthetic image from writing are allied to the visual percept.

This theory is certainly ingenious, and the experimental evidence seems to be unquestionable. It is supported by the work of Hincks (65) who found that several cases of reading disability were left-handed. Left-handed children are frequently encouraged or forced to write with the right hand; thus in writing the dominance of the left hemisphere is favoured, and some conflict with the naturally dominant right hemisphere is likely to occur. On the other hand, Smith (133) found that a confusion between 'b' and 'd', and 'p' and 'q' was frequent with normal readers when these letters were presented singly, but was much less marked when the letters were combined with others as nonsense syllables. Thus it is possible that reversal of letters and words may be a normal result of the child's inability to discriminate accurately minute differences of form and structure, and its tendency to perceive forms vaguely and generally. Finally one must be slow to support any hypothesis which depends upon the somewhat discredited theory of 'engrams'. Without resorting to a belief in this part of the hypothesis, however, one may recognize that conflict between the dominant and non-dominant hemispheres may be related to reading disability, as it is to other childish disabilities such as mirror writing and stammering.

Chapter VII

TYPOGRAPHICAL FACTORS

(1) Investigation of the Influence of Typographical Factors upon Reading

It is perhaps desirable to conclude with a survey of the influence of various typographical factors upon reading. It has been shown in previous chapters that the objective factors of the reading stimulus are of importance in providing a suitably structuralized and differentiated perceptual object. Thus it appears that letters and words should be sufficiently differentiated from one another to provide such an object. But the fact that we do not require to perceive individually every letter of every word has often led to the cry that our type forms are far too complicated, and that the eye is required to perform an increasingly difficult and fatiguing task in observing an immense number of small and minutely detailed symbols before the reader can apprehend even a moderate amount of the reading content. In this respect our method of spelling is as much at fault as our manner of writing and printing; and systems of reformed spelling have been adopted and are in use in the United States. Such systems, however, inevitably suppress the etymological relations between words, and between the English language and other languages. It is, then, a moot point whether the gain in ease and speed of reading is sufficiently great to offset the loss of interest and help afforded by these derivations and relationships.

Proposals for typographical reform have been almost as frequent as proposals for spelling reform, and in most cases have been framed upon an even slighter experimental basis. This subject is discussed in great detail by R. L. Pyke,[1] in his *Report* to the Medical Research Council on 'The Legibility of Print' (122). But a further discussion from a somewhat different aspect will not be out of place here. It should be stated that certain apparently fatal difficulties present themselves at the outset of any attempt to investigate the effects of typographical factors on reading. In the first place, two forms of experimental material can be utilized: (1) groups of letters, nonsense words or disconnected words, (2) connected and meaningful prose material. Now it is clear that the experimental situation presented by the first form of material is totally different from that presented by the second, because all the factors connected with the meaning and general significance of the content, the importance of which has been stressed again and again, are absent in the first case. Moreover the nature of the percept is quite different; for, as was noted before, the individual letters are not separately perceived in normal reading, as they must be in the reading of letters and nonsense words. And it is not necessary for individual words to be separately perceived in normal reading, as in the reading of disconnected words. Again it was

[1] Pyke also gives a detailed historical résumé of all the literature bearing on legibility. For details of sources of the results and opinions given below, the reader is referred to this *Report*, pp. 64–112.

shown that even the ocular motor habits are different in reading the two forms of material. Hence conclusions resulting from the use of material of the first kind are by no means valid when applied to normal reading. Indeed, Pyke found that there was no correlation between the relative legibility of various type faces as determined by the tachistoscopic reading of nonsense words, and of prose. If material of the second kind is employed, however, it is almost impossible to select passages such that the effects of differences of meaning, familiarity and secondary thought processes will not far outweigh the experimentally introduced typographical differences. Thus Hovde(69) has shown that the reading rate and legibility even of specially selected prose material is far more affected by differences in the context than by differences in the sensory content; for instance, in the size of type and leading. Gilliland(57) made his subjects read the same paragraph twice over, in different sizes of type. But this by no means makes the reading situation constant, for the subjective reactions of the reader will probably differ considerably at the first and second readings. Indeed Gilliland showed that the rate of reading and the number and duration of the fixations and their location in the line often varied greatly from the first reading to the second, showing that even the involuntary reactions were not constant.

Yet further difficulties present themselves. One is fully discussed by Pyke; namely, upon what measure shall the criterion of legibility be based? Probably the best ultimate measure would be the degree of fatigue produced by prolonged reading. But the experimental conditions are difficult to control, and, moreover, we have no definite accurate and extrinsic measure of fatigue, either ocular or general. Since, however, the crusade for typographical reform has been based, notably by Cohn(26) and Javal(75), on the unnecessary fatigue induced by reading, and the frequency of ocular defect which is stated to result from it, it would seem that an attempt at this method of measurement must be resorted to eventually. Moreover, the influence of variations in the reading performance set up by the meaning of the content would be minimized. In general the lateral eye movements do not seem to be greatly affected by fatigue. Dearborn(32) found that the accuracy of the long return movements from the end of one line to the beginning of the next was impaired by ocular fatigue, but his observation was made on only one subject, and has not been repeated and confirmed. It is quite possible that the power of accommodation would be more liable to fatigue than any other part of the visual apparatus, since accommodation varies throughout with the varying distance of different parts of the printed page from the eye. Unfortunately no satisfactory method of measuring the power of accommodation has been devised.

The most frequently employed criteria of legibility are speed of reading, accuracy of reading, and eye movements. Speed of reading can be modified enormously by the attitude of the reader. Errors in normal reading are very few; they are usually due either to some mistake in the assimilated meaning, or, in oral reading, to errors of vocalization. It was shown in Chap. IV (p. 50) that only very extreme typographical changes, such as the use of type more than 36 point or less than 6 point in size, or of a Gothic type face, affected the eye movements appreciably. The perceptual-motor reading habits are so well established that they are unaffected by moderate changes of the perceptual stimulus. And here another difficulty presents itself. It was shown by Pyke (122) that on the whole the mature reader found that the types to which he was the most accustomed were the most legible. Thus it would be difficult to assess the advantage of any typographical change, for in so far as it altered the perceptual situation, so far would it tend to decrease the legibility by rendering the situation unfamiliar, and the stimulus more difficult to perceive. Thus we are driven to the conclusion that it is almost impossible to assess legibility from the reading of the mature reader, unless some test of cumulative fatigue is ultimately devised.

The same limitations, however, need not apply to children's reading. In children, the perceptual and motor habits are less well established and more easily destroyed. Moreover, children pay more attention to the detailed structure of the perceptual stimulus, and may even perceive letters individually. For this reason, it is more important that the perceptual stimulus should be suitably structuralized, and its parts easy to discriminate. Also the eye of the child is much more liable to fatigue and strain than that of the adult. Hence it appears that the criterion of legibility should be based upon the reading of children. Type which is suitable for them will hold no difficulties for the adult, though certain modifications, such as a reduction in size, may be found desirable. For once the ocular motor habits have been formed and well established upon a hygienic basis, they will continue to function efficiently as long as the type is not greatly altered.

(2) General Typographical Standards

With these provisos we may turn to a consideration of the typographical standards which have actually been evolved. Huey(72) has summarized the findings of the earlier writers with regard to the size of type. These are based either upon the reading of adult subjects, or apparently merely upon opinion.

Height of short letters: 1·5 to 2·0 mm. (about 10 point).[1]

Width of letters: 6 or 7 per cm.

Thickness of vertical strokes: 0·25 to 0·30 mm.

Spaces within letters (between vertical strokes): 0·3 to 0·5 mm.

Spaces between letters: 0·50 to 0·75 mm.

Spaces between words: 2·0 mm.

It is a little difficult to see how such detailed figures are arrived at by mere guess-work. It is, of course, true that there must be an optimum size of type for adults, since if the print is very small, discrimination will be difficult, while if it is too large, each perceptual unit will contain only a small number of letters and assimilation will be retarded. However, the optimum size may cover a considerable range; and this view is

[1] See note on p. 50.

supported by the observations of Gilliland (57), who found that the rate of reading did not decrease greatly till the size was over 36 point or under 6 point. Hovde (69) found that a narrow variation of type size, from $6\frac{3}{4}$ to 8 point, had no significant effect upon the rate of reading. Paterson and Tinker (116), however, found that 10 point type was read faster than 6, 8, 12 or 14 point type; so that possibly 10 point type is the optimum size, because it is the most habitual. Blackhurst (12) found that the optimum sizes for children in the first, second, and third and fourth grades (aged approximately six, seven, and eight and nine years) were 24 point, 18 to 24 point, and 18 point respectively. It seems reasonable that children should require larger type than adults, since their span of perception and assimilation is narrower. On the other hand, Blackhurst's observations have not been confirmed by other workers. It would be interesting to have a complete and continuous series of observations with readers of all ages. It is possible that a rather small type, even if it produces rapid reading initially, may be more fatiguing eventually.

With regard to the length of line, there seems to be a consensus of opinion that it should not exceed 10 cm. Dearborn (32) and Huey (72) found that the number of fixations was fewer for shorter lines of print, but it does not appear that they were fewer proportionately to the number of words in the line. Dearborn, however, states that the return movements from the end of one line of print to the beginning of the next are likely to be more inaccurate the greater the length of line. This probably would occur with very unusually long lines, but there does not seem to be much evidence to show that such inaccuracy is greatly decreased with lines of less than 9 or 10 cm. in length. Blackhurst(12) found that the most suitable length of line for all the first four grades was 10 cm.; the rate of reading for this length of line was uniformly greater than the rate of reading for lines of 5·5, 8·0, 9·0, 12·0 and 14·0 cm. Tinker and Paterson(143a), however, found that adults reading prose paragraphs with lines 59, 80, 97, 114, 136, 152, 168 and 186 mm. in length respectively attained their maximum speed with lines of 80 mm. Paragraphs with lines of 97 mm. were read 2·8 per cent. more slowly, and this difference was not fully significant; but all other paragraphs were read more than 5 per cent. more slowly, and the differences were significant. Thus Tinker and Paterson concluded that the optimum length of line was from 75 to 90 mm. Hence we may state that the length of line should be not less than 7·5 cm., and not more than 10 cm.

Opinion seems to differ as to the importance of leading or interlineage. Javal (75) and Huey (72) state that interlineage should not be used at the expense of body size of type; that is, the type should not be decreased below a certain size, about 10 point, in order to allow room for interlineage. But the work of certain investigators (Madison Bentley (9), Griffing and Franz (61)) seems to show that leading is an advantage with small type. Baird (4) found that the use of ½ point lead with small type increased the speed at which numbers could be found in a telephone directory. Opinion on the whole is favourable to the use of leading with larger sizes of type. The most detailed work on the subject was done by Madison Bentley(9), with adult readers reading at top speed. He found that speed seemed to depend upon the ratio—letter height/interlinear space; thus the optimum interlineage was:

Approximately $\frac{1}{10}$ in. for 12 point type,
" $\frac{1}{14}$ in. " 9 "
" $\frac{1}{17}$ in. " 6 "

Blackhurst(12) obtained inconclusive results with regard to the importance of leading; these did not show that any advantage accrued from the use of more than 1 point leading, that is, an interlineage of 1·33 mm. A printed page always appears clearer and more legible with a moderate amount of interlineage. For normal reading, however, it is probable that an adequate size of type is more important.

Even less is known with certainty as to the desirability of marginal indentation, that is, an alternation of lines of irregular length. Irregular indentation, such as is produced by the inclusion of marginal figures in the text, would certainly appear to be undesirable, since, according to Dearborn(32), it upsets the 'short-lived motor habits'; that is, it interferes with the habitual length of the return movements. On the other hand, it has been suggested that a slight indentation at the left-hand end of alternate lines would not interfere with regular habits, and would help to prevent a return to the line above or the line below the correct one. It seems quite possible that children have a tendency to return to the wrong line.

Indeed Blackhurst(12) found that first-grade children read more rapidly and correctly printed material in which the left-hand margin was irregular, but he does not state the nature of the irregularity. He questions, however, if it is not most desirable that the child should become accustomed from the start to the regular left-hand margin. Certainly this is true if it is to be obliged ultimately to use a regular left-hand margin. However, it might be advantageous to use the indented margin for adult readers. It appeared in some work of the writer's (147) that readers did occasionally jump or recede a line in returning from the end of one line to the beginning of the next; though this occurred only once or twice, on the average, in ten pages of reading. There is perhaps insufficient evidence for so radical a change in the layout of the printed page.

No experimental evidence is available as to the best width of margin. An attempt was made by the writer to find the effects upon speed of reading of changes in the width of margin, but the results were quite inconclusive. Even the preferences expressed by the subjects of the experiments showed nothing. Most readers did not notice the changes of marginal width, and those who did were mutually contradictory in their preferences. The printer, of course, regards the double page as a unit, and considers that the double block of printed material should be arranged upon the page as a picture is arranged upon a mount. Thus the inside margin of the single page becomes the narrowest, and then the top and then the outside; the bottom margin is the widest. It seems probable that the width of the upper, lower and outside margins are unimportant except from the aesthetic point of view. But the use of a narrow inner margin may be undesirable, especially in thick, closely bound books, in which the inner part of the print becomes bent over, and the two blocks of print lie very near to one another. Thus it should be a rule that every book should be printed and bound in such a way that all the printed part of the page lies flat, and that there is an adequate marginal space between the two blocks of print, even in the middle of the book.

A variety of opinion has been expressed as to the best colour for the paper. Javal(75) and Weber (151) held that white paper was too glaring and suggested yellowish, but Cohn (26) stated that yellowish paper produced bluish after-images. A compromise of cream colour has also been suggested, but the general consensus of opinion is in favour of white. The surface should, however, be matt, not glossy, to give diffuse and avoid specular reflection. The paper should of course be thick enough to prevent the ink from showing through it. From experimental evidence, Griffing and Franz(61) found that white, greyish, yellow and red paper were best in order of legibility. Starch(134) quotes from an unpublished paper by Hall and Ames, who found that matt white, medium glossy pink and blue, and glossy white, in order, gave the best legibility. Black ink is generally considered to be the best.

(3) Forms of Letters

As stated above, it has frequently been suggested that it would be desirable to alter the form of some of the individual letters which are very similar to one another in shape. Thus at a glance one can see that 'o' and 'e', 'a' and 's', 'r', 'n' and 'u', 'f' and 't', 'i' and 'l', 'b', 'h' and 'd' are only slightly differentiated from each other. Numerous tachistoscopic studies of the recognition of single letters have amply supported these conclusions; a useful summary of them is given by Tinker(143). But in the light of the qualifications adduced on p. 161 above, it may safely be stated that these results cannot be held to apply to normal reading, unless they are confirmed by similar results obtained from normal reading. Now in a detailed study by the writer(147) it was found that, in the tachistoscopic reading of adults, the number of letters which were mistaken for other letters of similar form decreased when short sentences were substituted for disconnected words, and longer sentences for short ones. The number of errors which depended purely on confusion of appearance, that is to say, where the substituted word had quite a different meaning from the word it replaced although resembling it in form, was only 4 per 1000 words for educated readers reading the longer sentences. For another group of less educated readers with rather more difficult material, it varied from 9 to 23 per 10 pages. Of these errors, less than half were the result of pure substitution of one or more letters; the remainder involved also the insertion or, more commonly, the omission of letters. This omission of letters was much more frequent among the less educated than among the more educated readers. It was shown that there was a tendency among the former, when they did not fully understand what they were reading, to replace long and unfamiliar words by shorter and more familiar ones of a similar form. There was no evidence to show that a reader who understood what he was reading, and was fully cognisant of its context, was ever liable to mis-read words purely on account of the ambiguous and indistinguishable form of their letters. The chances are that in any word there will be a sufficient number of clearly perceived letters to produce an unambiguous perceptual object, provided that sufficient meaning exists to ensure its interpretation. If such meaning does not exist, either another more familiar word of similar appearance will be substituted; or, more probably, when the reader is not hurried as he is in tachistoscopic reading, the word will be re-examined. If he resorts to a detailed scrutiny, it seems likely that he will observe the letters sufficiently closely to obviate any letter confusion. Thus it is only the child who is beginning to read who will be liable to confuse actual letters.

It seems to be generally considered that children do tend to confuse words of similar form, although either the total word form, or certain parts of it, may be responsible for the confusion. Meek (99) found that children used certain letters or groups of letters as cues for recognition; and if these letters were misread, it seems probable that the child would mis-read the whole word. But N. B. Smith (133) found that letters which were easily confused when presented singly were much less liable to be confused when presented in combination as nonsense syllables. The conclusion is that only words with a number of ambiguous letters would be mis-read simply as a result of letter confusion. But unfortunately there seems to be definite evidence on this point.

The letters which appeared to be liable to mutual confusion were mentioned above. In the writer's work(147) it was found that the most common form of definite letter-for-letter substitution which caused a mis-reading was a confusion of the ascending letters, particularly of 'f' and 't', and 'l' and 't'. Confusion of vowels was also frequent, not only of 'o' and 'e', but also of 'a' and 'e'. These substitutions are, of course, determined not only by similarity of letter form, but also by the existence of words which are differentiated from one another by one or two letters only; and these differentiating letters are frequently the vowels 'a', 'e', and 'o', and also the ascending letters 'l' and 't' (less commonly 'f'). These results seem to indicate that if letter confusion occurs in children's reading, it is likely to lead to the substitution of words differentiated from the correct ones only by the common vowels or the common ascenders. This is the most which can be said definitely with regard to ambiguous letter forms. Further adequately controlled experimental evidence is required before we can do more than surmise the liability of children to confuse these letters.

(4) Type Faces

Finally, it is necessary to consider the question of type face. Here the issue has been somewhat confused by the rivalry between the claims of legibility and of aesthetic experience. For a full and historical treatment of the aesthetic factor the reader is referred to the works of Stanley Morison, and to *Printing Types*, by D. B. Updike. These writers recommend a simple, fairly broad type face, without great contrast between the thick and thin strokes. Thus simple Old Face types, such as Caslon, are to be preferred to Modern Face types. There is little doubt that aesthetic opinion nowadays is in favour of Old Style or Old Face types; these are employed for the majority of well-printed books. This may, however, be merely a matter of fashion, similar to the fashion for Baskerville and Bodoni Modern Faces in the last century. It has this practical justification, that the very fine serifs[1] and hair lines of a Modern Face type are liable to get broken, or to print off so badly as to be almost invisible; hence in part the illegibility of linotype newspaper print.

Javal(75) recommended that Old Style type should be used for children; since the letters are of fairly uniform thickness they are more legible for children who tend to read individual letters. For adults, however, Modern Styles are preferable, since the total word form is more differentiated, by the alternation of main strokes and hair lines. But the long narrow serifs of Modern Faces tend to run into one another,

[1] The serifs are the short horizontal strokes which define the extremities of the vertical strokes of the letters.

and the Old Style short triangular serifs are more legible and durable. Javal, indeed, designed a type face of his own which was calculated to produce the maximum legibility of all the letters. But as previously remarked, his conclusions were not based upon experimental evidence, and they have been assailed by later writers. The results of later experimental work must be almost entirely discounted, since the material and conditions were different in every way from those encountered in normal reading. Pyke's results (122) are inconclusive, though more reliable; he found that, on the whole, an Old Style type face was the most legible. He concluded, on the basis of these and of previous results and dicta, that the ideal type 'should be simple, fairly broad, with fairly thick limbs, but not too much contrast in thickness and thinness, and with fairly wide spacing.' Thus there seems to be a general bias in favour of the employment of Old Style or Old Face types. But Javal's theory in favour of the use of Modern Face types for adults has never been either proved or disproved. This point could, in all probability, only be decided upon the basis of the fatigue criterion, as described above.

Thus we see that not much is known definitely as to the influence of typographical factors upon the ease and efficiency of reading. As long as the existent methods of experimentation are employed, it seems unlikely that conclusions of importance will be reached.

CONCLUSION

In the preliminary treatment of movements of the eyeball it was shown that voluntary and reflex movements about the field of vision were in general inaccurate and erratic. To pass from one part of the field to another it was usually necessary to execute corrective movements interspersed with fixation pauses during which the eye reviewed its position. Nor could the eye be maintained steadily in one fixation position for any length of time; it was liable to move away, as the result of reflex stimulation from peripheral visual sensations. The movements of the two eyes were not always well integrated, but differed from each other both in speed and direction of movement. This behaviour still continued to a greater or less degree in the reading of the young child. Fixation pauses and movements to and fro along the printed lines were erratic and very numerous. But in mature adult reading, a series of regular and rhythmical backwards and forwards movements had developed, the forward movements alternating regularly with fixation pauses of comparatively short and regular duration. These movements were habitual, and could be carried on automatically in the absence of any attention to perception and assimilation of the reading content. It was shown that efficient motor habits could be developed not only in the normal reading of English, but also in the reading of foreign languages, disconnected words, algebraic and chemical formulae, and figures arranged in horizontal or vertical lines; these series of movements and fixation pauses were different in each case, but were equally well adapted for dealing with the material to be read. But they were liable to be abrogated by any species of mental struggle or conflict, whether in an effort to understand, or to perform some difficult task such as paraphrasing, translating or analysing the text, or as a result of an affective reaction to the reading content. In such cases 'periods of confusion' might supervene, during which there was a return to the irregular wandering eye movements of childish reading.

Passing to the perceptual processes in reading, it was shown that perception in children was no better adapted by nature for efficient reading than were the motor processes. Children's perceptions were as a rule vague, inaccurate and undifferentiated, and there was frequently a strong tendency to interpret the primary sensations in accordance with the general features of the total situation—that is, the perceptual setting, the meaning, familiarity and interest of the perceptual object, and a variety of purely subjective factors such as moods and attitudes. It appeared that there was no definite experimental evidence as to how perception developed with sufficient detail and accuracy for reading purposes. In adult reading, in general only a small part of the visual field was actually perceived, probably only the contours of the words together with a few prominent details; these were sufficient to suggest the corresponding language and thought units, and hence to convey the meaning of the content. It seemed to depend upon the method of teaching reading whether the child learnt gradually to differentiate and observe the important features of the vague general percept, or to build up a series of discrete unitary percepts into a total structure closely integrated with the corresponding language structures. However that might be, there was no doubt as to the importance throughout of preserving and stressing the background of familiarity, interest and associated thought. It was a necessary accompaniment of the child's first attempts at reading, and was of the utmost importance in filling out the vague partial perceptions of adult reading to produce assimilation of the general meaning of the content. For the same reason, once reading had become fully matured, details in the objective stimulus such as the size and nature of the printing type were usually ignored; the perceptual processes were so readily supplemented from the general trend of associated thought that a mere skeleton of the normal perceptual structure could be successfully assimilated. The printed or written words had become comparable to a primary sensory configuration or 'gestalt', in that they could receive any alteration except to their essential structure and yet be assimilated unchanged.

Thus it appeared that in mature adult reading motor and perceptual processes had been developed and adapted from their original and natural form to produce a well-integrated and highly efficient series of habitual responses.

REFERENCES

(1) ABERNETHY, E. M. *J. Educ. Psychol.* (1929), **20**, 695.
(2) AHRENS, A. *Die Bewegung der Augen beim Schreiben.* Rostock, 1891.
(3) ANDERSON, C. J. and MERTON, E. *Elemen. School J.* (1921), **21**, 336.
(4) BAIRD, J. W. *J. Appl. Psychol.* (1917), **1**, 30.
(5) BARANY, R. *Zsch. f. Sinnesphysiol.* (1911), **45**, 59.
(6) BARNES, B. *Amer. J. Psychol.* (1905), **16**, 199.
(7) BARTLETT, F. C. *Brit. J. Psychol.* (1916), **8**, 222.
(8) —— *Rep. of the Brit. Assoc. for the Advancement of Science* (1929), 186.
(9) BENTLEY, MADISON. *Psychol. Monog.* (1921), **30**, No. 3.
(10) BINET, A. *Année Psychol.* (1897), **3**, 296.
(11) —— *Ibid.* (1908), **14**, 1.
(12) BLACKHURST, J. H. *Investigations in the Hygiene of Reading.* Baltimore, Warwick and York, Inc. 1927.
(13) BLIX, M. *Upsala Lakärefs Förh.* (1890), **15**, 349.
(14) BOGGS, L. P. *Pedagogical Seminary* (1905), **12**, 496.
(15) BOWDEN, J. H. *Elemen. School J.* (1911), **12**, 21.
(16) BRAMWELL, E. *Brain* (1928), **51**, 1.
(17) BRONNER, A. F. *Psychology of Special Abilities and Disabilities.* London: Kegan Paul, Trench, Trübner and Co. Ltd. 1919.
(18) BROWN, H. A. *J. Educ. Res.* (1920), **2**, 436.
(19) BUSWELL, G. T. *Suppl. Educ. Monog.* (1920), No. 17.
(20) —— *Ibid.* (1922), No. 21.
(21) —— *Ibid.* (1926), No. 30.
(22) —— *A Laboratory Study of the Reading of Modern Foreign Languages.* New York: The Macmillan Co. 1927.
(23) CAMERON, E. H. and STEELE, W. M. *Psychol. Monog.* (1905), **7**, No. 1.
(24) CATTELL, J. McK. *Brain* (1886), **8**, 295.
(25) COBURN, E. B. *Arch. Ophth.* (1905), **34**, 1.
(26) COHN, H. *Rev. scient.* (1881), **27**, 290.
(27) CORDS, R. *Arch. f. Ophth.* (1927), **118**, 771.
(28) CURTIS, H. S. *Amer. J. Psychol.* (1899), **11**, 237.
(29) DALLENBACH, K. M. *J. Educ. Psychol.* (1914), **5**, 387.
(30) —— *Ibid.* (1919), **10**, 61.
(31) DEARBORN, W. F. *Psychol. Rev.* (1904), **11**, 297.
(32) —— *Arch. Philos. Psychol. Scient. Meth.* (1906), No. 4.

(33) DELABARRE, E. B. *Amer. J. Psychol.* (1898), **9**, 572.
(34) DESCOEUDRES, A. *Arch. de psychol.* (1914), **14**, 305.
(35) DIEFENDORF, A. R. and DODGE, R. *Brain* (1908), **31**, 451.
(36) DICKINSON, C. A. *Amer. J. Psychol.* (1926), **37**, 330.
(37) DODGE, R. and CLINE, T. S. *Psychol. Rev.* (1901), **8**, 145.
(38) DODGE, R. *Amer. J. Physiol.* (1903), **8**, 307.
(39) —— *Psychol. Bull.* (1905), **2**, 55.
(40) —— *Ibid.* (1905), **2**, 193.
(41) —— *Psychol. Monog.* (1907), **8**, No. 4.
(42) —— *J. Exper. Psychol.* (1921), **4**, 165.
(43) —— *Ibid.* (1921), **4**, 247.
(44) —— *Ibid.* (1923), **6**, 1.
(45) ENGELKING. *Klin. Monatsbl. f. Augenhk.* (1922), **68**, 53.
(46) ERDMANN, B. and DODGE, R. *Psychologische Untersuchungen über das Lesen, auf experimenteller Grundlage.* Halle: 1898.
(47) FERNALD, G. M. and KELLER, H. *J. Educ. Res.* (1921), **4**, 355.
(48) FOSTER, W. S. *J. Educ. Psychol.* (1911), **2**, 11.
(49) FREEMAN, F. N. *J. Appl. Psychol.* (1920), **4**, 126.
(50) FREEMAN, G. L. *J. Exper. Psychol.* (1929), **12**, 340.
(51) GALLEY, L. *Zsch. f. Psychol.* (1927), **101**, 182.
(52) GATES, A. I. *Teachers' Coll. Contrib. to Educ.* (1922), No. 129.
(53) —— *J. Educ. Psychol.* (1926), **17**, 289.
(54) —— *Ibid.* (1927), **18**, 217.
(55) GATES, A. I. and BOEKER, E. *Teachers' Coll. Record* (1923), **24**, 469.
(56) GILLILAND, A. R. Unpublished thesis. Abstracted in *Psychol. Bull.* (1922), **24**, 622.
(57) —— *Elemen. School J.* (1923), **24**, 138.
(58) GOLDSCHEIDER, A. and MÜLLER, E. *Zsch. f. klin. Med.* (1893), **23**, 131.
(59) GRAY, C. T. *Suppl. Educ. Monog.* (1917), **1**, No. 5.
(60) GRAY, W. S. *Elemen. School J.* (1921), **21**, 577.
(61) GRIFFING, H. and FRANZ, S. I. *Psychol. Rev.* (1896), **3**, 513.
(62) GRINDLEY, G. C. Unpublished work.
(63) GUILLERY. *Arch. f. d. ges. Physiol.* (1898), **73**, 87.
(64) HAMILTON, F. M. *Arch. Psychol.* (1907), No. 9.
(64*a*) HEIMANN, A. and THORNER, H. *Arch. f. d. ges. Psychol.* (1929), **71**, 165.
(65) HINCKS, E. M. *Harvard Monog. in Educ.* (1926), **2**, No. 2.
(66) HINSHELWOOD, J. *Congenital Word Blindness.* London: H. K. Lewis and Co. 1917.
(67) HOLLINGWORTH, L. S. *Special Talents and Defects.* New York: The Macmillan Co. 1923.
(68) HOLT, E. B. *Psychol. Monog.* (1903), **4**, No. 1.
(69) HOVDE, H. T. *J. Appl. Psychol.* (1930), **14**, 63.
(70) HUEY, E. B. *Amer. J. Psychol.* (1898), **9**, 575.

(71) HUEY, E. B. *Amer. J. Psychol.* (1901), **11**, 283.
(72) —— *The Psychology and Pedagogy of Reading.* New York: The Macmillan Co. 1910.
(73) JAENSCH, E. R. *Eidetic Imagery.* London: Kegan Paul, Trench, Trübner and Co. Ltd. 1930.
(74) JAVAL, E. *Année d'Ocul.* (1878), **79**, 197 and 240.
(75) —— *Physiologie de la Lecture et de l'Écriture.* Paris: Félix Alcan, 1905.
(76) JUDD, C. H. and COURTEN, H. C. *Psychol. Monog.* (1905), **7**, No. 1.
(77) JUDD, C. H., MCALLISTER, C. N. and STEELE, W. M. *Ibid.* (1905), **7**, No. 1.
(78) JUDD, C. H. *Psychol. Monog.* (1905), **7**, No. 1.
(79) —— *Ibid.* (1907), **8**, No. 3.
(80) —— *J. Philos. Psychol. Scient. Meth.* (1909), **6**, 36.
(81) —— *Suppl. Educ. Monog.* (1918), **2**, No. 4.
(82) JUDD, C. H. and BUSWELL, G. T. *Ibid.* (1922), No. 23.
(83) KOCH, E. *Arch. f. d. ges. Psychol.* (1908), **13**, 196.
(84) KOFFKA, K. *Psychol. Bull.* (1922), **19**, 531.
(85) —— *The Growth of the Mind.* London: Kegan Paul, Trench, Trübner and Co. Ltd. 1928.
(86) KÖHLER, W. *Gestalt Psychology.* London: G. Bell and Sons, Ltd. 1930.
(87) KOLEN, A. A. *Arch. f. Augenhk.* (1926), **97**, 341.
(88) KORTE, W. *Zsch. f. Psychol.* (1923), **93**, 17.
(89) KUTZNER, O. *Arch. f. d. ges. Psychol.* (1916), **35**, 157.
(90) LAMANSKY, S. *Arch. f. d. ges. Physiol.* (1869), **2**, 418.
(91) LAMARE. *C. r. à Soc. franç. d'opht.* (1892), **10**, 354.
(92) LEIRI, F. *Arch. f. Ophth.* (1927), **119**, 711.
(92*a*) LINE, W. *Brit. J. Psychol. Monog.* (1931), No. 15.
(93) LORD, E. E., CARMICHAEL, L. and DEARBORN, W. F. *Harvard Monog. in Educ.* (1925), No. 6.
(94) LORING, M. *Psychol. Rev.* (1915), **22**, 354.
(95) MCALLISTER, C. N. *Psychol. Monog.* (1905), **7**, No. 1.
(96) MACDOUGALL, R. *Amer. J. Physiol.* (1903), **9**, 122.
(97) —— *Ibid.* (1915), **37**, 300.
(98) MARX, E. and TRENDELENBURG, W. *Zsch. f. Sinnesphysiol.* (1911), **45**, 87.
(99) MEEK, L. H. *Teachers' Coll. Contrib. to Educ.* (1925), No. 164.
(100) MESSMER, O. *Arch. f. d. ges. Psychol.* (1904), **2**, 190.
(101) MILES, W. R. and SHEN, E. *J. Exper. Psychol.* (1925), **8**, 344.
(102) MILES, W. R. *J. Gen. Psychol.* (1928), **1**, 373.
(103) —— *Psychol. Rev.* (1929), **36**, 122.
(104) MONROE, M. *Genetic Psychol. Monog.* (1928), **4**, No. 4.
(104*a*) MOSHER, R. M. *J. Educ. Psychol.* (1928), **19**, 185.
(105) NEWHALL, S. M. *Amer. J. Psychol.* (1928), **40**, 628.

(106) O'Brien, J. A. *Silent Reading*. New York: The Macmillan Co. 1921.
(107) Ohm, J. *Zsch. f. Augenhk.* (1914), **32**, 4.
(108) —— *Arch. f. Ophth.* (1928), **120**, 233.
(109) Öhrwall, H. *Skand. Arch. f. Physiol.* (1912), **27**, 33.
(110) Orchansky, J. *Centbl. f. Physiol.* (1898), **12**, 785.
(111) Orton, S. T. *J. Amer. Med. Ass.* (1928), **90**, 1095.
(112) —— *J. Educ. Psychol.* (1929), **20**, 135.
(113) Parsons, C. J. *Brit. J. Psychol.* (1917), **9**, 74.
(114) Parsons, J. H. *Colour Vision*. Cambridge University Press, 1915.
(115) —— *An Introduction to the Theory of Perception*. Cambridge University Press, 1927.
(116) Paterson, D. G. and Tinker, M. A. *J. Appl. Psychol.* (1929), **13**, 120.
(117) Piaget, J. and Rossello, P. *Arch. de psychol.* (1922), p. 208.
(118) Pillsbury, W. B. *Amer. J. Psychol.* (1897), **8**, 315.
(119) Piltz, J. *Neurol. Centbl.* (1904), **23**, 801.
(120) Pintner, R. *J. Educ. Psychol.* (1913), **4**, 333.
(121) —— *Psychol. Rev.* (1913), **20**, 129.
(122) Pyke, R. L. *Med. Res. Counc. Spec. Rep. Series* (1926), No. 110.
(122*a*) Pyle, W. H. *Elemen. School J.* (1929), **29**, 597.
(123) Quantz, J. O. *Psychol. Monog.* (1897), **2**, No. 1.
(124) Rogers, A. S. *Amer. J. Psychol.* (1917), **28**, 519.
(125) Rubin, E. *Visuell wahrgenommen Figuren. Studien in psychologischer Analyse*. Kobenhavn: Gyldendal, 1921.
(126) Ruediger, W. C. *Arch. Psychol.* (1907), No. 5.
(127) Schackwitz, A. *Zsch. f. Psychol.* (1913) **63**, 442.
(128) Schmidt, W. A. *Suppl. Educ. Monog.* (1917), **1**, No. 2.
(129) Secor, W. B. *Amer. J. Psychol.* (1899), **11**, 225.
(130) Segers, J. E. *J. de psychol.* (1926), **23**, 608 and 723.
(131) Sholty, M. *Elemen. School J.* (1911), **12**, 272.
(132) Smith, F. *Brit. J. Psychol.* (1914), **6**, 321.
(133) Smith, N. B. *J. Educ. Psychol.* (1928), **19**, 560.
(134) Starch, D. *Educational Psychology*. New York: The Macmillan Co. 1919.
(134*a*) Stein, W. *Arch. f. d. ges. Psychol.* (1928), **64**, 301.
(135) Stern, W. *Psychology of Early Childhood*. London: George Allen and Unwin, Ltd. 1924.
(136) Stevens, G. T. *A Treatise on the Motor Apparatus of the Eyes*. Philadelphia: F. A. Davis and Co. 1906.
(137) Stratton, G. M. *Wundt's Phil. Stud.* (1902), **20**, 336.
(138) —— *Psychol. Rev.* (1906), **13**, 81.
(139) Struycken, J. L. *Ned. Tijdschr. v. Gen.* (1918), **1**. Abstracted in *Zsch. f. Psychol.* (1918), **85**, 345.
(140) Sundberg. *Skand. Arch. f. Physiol.* (1917), **35**, 1.
(141) Terry, P. W. *Suppl. Educ. Monog.* (1922), No. 18.

(141 *a*) THORNER, H. *Arch. f. d. ges. Psychol.* (1929), **13**, 185.
(142) TINKER, M. A. *Genetic Psychol. Monog.* (1928), **3**, No. 2.
(143) —— *J. Gen. Psychol.* (1928), **1**, 472.
(143 *a*) TINKER, M. A. and PATERSON, D. G. *J. Appl. Psychol.* (1929), **13**, 205.
(144) TOTTEN, E. *J. Comp. Psychol.* (1926), **6**, 287.
(145) VERNON, M. D. *Brit. J. Ophth.* (1928), **12**, 113.
(146) —— *Brit. J. Psychol.* (1929), **20**, 161.
(147) —— *Med. Res. Counc. Spec. Rep. Series* (1929), No. 130, Part A.
(148) —— *Brit. J. Psychol.* (1930), **21**, 64.
(149) —— *Med. Res. Counc. Spec. Rep. Series* (1930), No. 148.
(150) —— *Brit. J. Psychol.* (1931), **21**, 368.
(151) WEBER, A. *Die Augenuntersuchungen in den höheren Schulen zu Darmstadt. Referat und Memorial, erstattet der grossherzogliche Ministerial-Abtheilung für Gesundheitspflege.* Leipzig: H. Brill, 1881.
(152) WEISS, O. *Zsch. f. Sinnesphysiol.* (1911), **45**, 313.
(153) WHIPPLE, G. M. *J. Educ. Psychol.* (1910), **1**, 249.
(154) WHIPPLE, G. M. and CURTIS, J. N. *Ibid.* (1917), **8**, 333.
(155) WIEDERSHEIM, O. *Klin. Monatsbl. f. Augenhk.* (1928), **80**, 380.
(156) WILEY, W. E. *J. Educ. Res.* (1928), **17**, 278.
(156 *a*) WILSON, E. *Elemen. School J.* (1922), **22**, 380.
(157) ZEITLER, J. *Wundt's Phil. Stud.* (1900), **16**, 380.

INDEX

Printed in the United States
By Bookmasters